539.38 : 621.753

STRAIN GAUGE TECHNOLOGY

STRAIN GAUGE TECHNOLOGY

Edited by

A. L. WINDOW

Welwyn Strain Measurement Ltd,
Basingstoke, Hampshire, UK

and

G. S. HOLISTER

Faculty of Technology, The Open University,
Walton Hall, Milton Keynes, UK

APPLIED SCIENCE PUBLISHERS
LONDON and NEW JERSEY

APPLIED SCIENCE PUBLISHERS LTD
Ripple Road, Barking, Essex, England
APPLIED SCIENCE PUBLISHERS INC.
Englewood, New Jersey 07631, USA

British Library Cataloguing in Publication Data

Strain gauge technology.
1. Strain gages
I. Window, A. L. II. Holister, G. S.
620.1′123028   TA413.5

ISBN 0-85334-118-4

WITH 30 TABLES AND 142 ILLUSTRATIONS

© APPLIED SCIENCE PUBLISHERS LTD 1982

Printed in Great Britain by Galliard (Printers) Ltd, Great Yarmouth

PREFACE

Experimental stress analysis is the strictly practical branch of stress analysis and for most problems it relies heavily on strain measurement techniques. Of these, by far the most widely used is that involving strain gauges in general and electrical resistance strain gauges in particular. These small and basically simple devices have been in use for many years, not only in all branches of engineering, but in almost all industries, and crossing all disciplines. In spite of, or perhaps because of, their enormous field of application, there have been very few published books on the subject. Most of the literature consists of technical papers and reports, referring to specific problems or particular industries.

Strain Gauge Technology, which grew out of Applied Science Publishers' *Developments in Stress Analysis* series, does not replace or supersede existing text books and reference books. The basic theory of strain gauges has not changed, but the performance now available, and the techniques of installation, calibration, and data capture are very different to those used even a few years ago. There have been no revolutionary inventions; what has occurred is a very important steady improvement in materials, and in the techniques of both manufacture and installation, which has made strain gauges far more reliable, accurate, and versatile. They should now be the everyday tools of all engineers concerned with design, prototype testing, or development, as well as for failure analysis.

The performance of precision transducers such as load cells and electronic weighing scales is now very much accepted and even taken for granted. It is often forgotten, or perhaps not known, that the sensors in these devices are usually resistance strain gauges. With commercial scales used in retail shops already offering accuracies of 0.03% and approaching

v

0.02% here is a true indication of the performance and reliability available from resistance strain gauges.

In Part I, Electrical Resistance Strain Gauges, Chalmers discusses present day materials and construction methods for gauges and their performance characteristics. Mordan describes the various adhesives used and the installation methods for a wide variety of applications and performance requirements. This subject is taken further by Pople to cover the particular problems of installation and protection in difficult and hostile environments. Scott and Owens discuss strain gauge instrumentation at length, reviewing the well known and traditional equipment, but adding a powerful introduction to computerised data acquisition and reduction, and to the trends towards high speed digital systems. Part I is concluded with a unique exposition on error analysis which should give every strain gauge practitioner cause for thought.

Part II covers three very important though not so widely used strain gauge groups. Baker introduces and updates the performance characteristics of semiconductor or piezo resistive strain gauges. Procter and Strong review the available capacitance strain gauges, increasingly used for static strain measurement at elevated temperature, and Hornby describes and discusses applications of vibrating wire strain gauges which are particularly well suited to measuring long-term static strains in civil engineering.

This book is intended to fill an important gap between the manufacturers' literature which gives detailed instructions for handling a particular product, and a text book which covers basic theory and general principles of strain measurement. It brings together much needed practical 'know-how' and experience from a number of leading exponents of the various aspects of strain gauge practice. All the authors are practising engineers in the real commercial world, and whilst some are well known for their published work, others have been persuaded for the first time to commit their knowledge and experience to paper. It is hoped that the result will prove equally useful to both 'do-ers' and 'specify-ers.'

A. L. WINDOW
G. S. HOLISTER

CONTENTS

LIST OF CONTRIBUTORS

M. A. BAKER, C.Eng., M.I.Mech.E.

Technical Director, Maywood Instruments Ltd, Whitney Road, Daneshill Industrial Estate, Basingstoke, Hampshire RG24 0NS, UK.

G. F. CHALMERS, C.Eng., M.R.Ae.S.

Vice-President, Director of International Marketing, Measurements Group Inc., P.O. Box 27777, Raleigh, North Carolina 27611, USA.

I. W. HORNBY, C.Eng., M.I.C.E.

Head of Concrete Technology Group, Central Electricity Research Laboratories, Kelvin Avenue, Leatherhead, Surrey KT22 7SE, UK.

G. C. MORDAN

Customer Service Manager, Welwyn Strain Measurement Ltd, Armstrong Road, Basingstoke, Hampshire RG24 0QA, UK.

A. OWENS, B.Eng., Ph.D.

Analytical Services Manager, Stress Engineering Services Ltd, Charlton Lane, Midsomer Norton, Bath BA3 4BD Avon, UK.

J. POPLE, M.Inst.M.C.

Deputy Head, Instrumental Laboratory, Vickers Shipbuilding and Engineering Ltd, P.O. Box 6, Barrow-in-Furness, Cumbria LA14 1AB, UK.

E. PROCTER, C.Eng., M.I.Mech.E.

Head of Experimental Structural Analysis Section, Central Electricity Generating Board, Berkeley Nuclear Laboratories, Berkeley, Gloucestershire GL13 9PB, UK.

K. SCOTT

Technical Representative, Welwyn Strain Measurement Ltd, Armstrong Road, Basingstoke, Hampshire RG24 0QA, UK.

J. T. STRONG, C.Eng., M.I.Mech.E.

Research Officer, Central Electricity Generating Board, Berkeley Nuclear Laboratories, Berkeley, Gloucestershire GL13 9PB, UK.

PART I

Electrical Resistance Strain Gauges

Chapter 1

MATERIALS, CONSTRUCTION, PERFORMANCE AND CHARACTERISTICS

G. F. CHALMERS

Measurements Group Inc., Raleigh, USA

INTRODUCTION

Since 1940, the bonded electrical resistance strain gauge has been the most powerful single tool in the field of experimental stress analysis. It is one of the most accurate, sensitive, versatile and easy-to-use sensors available. It is relatively low in cost, is linear in output, is easily installed, and is available in a broad variety of configurations, sizes and materials, to meet a very large spectrum of measurement requirements over a wide range of temperatures.

Continuous commercial development, and extensive industrial and research investigations, have resulted in excellent performance character-istics—particularly in stability, temperature compensation, and creep. This, in turn, has led to the strain gauge becoming the basic sensing element of very high precision load transducers and weighing systems, in which there have been rapid developments in recent years (Fig. 1).

In spite of the relative ease with which strain gauges can be employed, the proper and effective use of them requires a thorough understanding of their characteristics and performance, and of the application techniques and associated instrumentation. This chapter is devoted to those features which affect the behaviour of the resistance strain gauge itself. Later chapters deal with adhesives, installation and protection techniques, instrumentation, and errors.

1

FIG. 1. Foil strain gauge installation on precision transducer. Courtesy of FFA Sweden.

BRIEF HISTORY

The bonded electrical resistance strain gauge in the basic form known today was first used in the USA in 1938. Simmons at the California Institute of Technology, and Ruge at the Massachusetts Institute of Technology, both discovered that small diameter wires made of electrical resistance alloys, such as copper–nickel, could be adhesively bonded to a structure to measure surface strain.

The strain sensitivity of resistance wires was first utilised some years prior to this, however, in the unbonded wire strain gauge. This consisted of an arrangement of wire, wound around a series of pins actuated by linkages, any movement of which stretched the wire and changed its resistance. This was essentially an electrical extensometer, and the principle is still used today in some special types of transducers.

Much earlier still, it was Lord Kelvin who, in 1856, first reported on the relationship between strain and the resistance of wire conductors. He discovered that differences in tension of the wires he was using affected his resistance measurements. This was an inconvenience to him, and he carefully tried to minimise the effect.

It is also important to acknowledge the first use of a bonded resistance strain gauge device in the early 1930s. This consisted of a carbon composition resistor on an insulating strip, used to measure vibratory strains in high performance propeller blades for aircraft. The carbon gauge was, however, limited to the indication of dynamic strains, and could measure these to only a moderate degree of accuracy, owing to the lack of resistance stability with time and temperature.

Following the development of the techniques for bonding resistance wires to structures, discovered by Simmons and Ruge, strain gauge measurements were quickly adopted for structural testing during the rapidly growing aircraft development programmes of World War II. They continued to be used largely in the aircraft industry for a number of years, and it was the requirements of this industry that led to the significantly important development of the foil type of strain gauge in 1952. The Saunders–Roe Company in the UK was seeking improvements in the bonded wire gauges, which were presenting problems in harsh testing environments on helicopters and flying boats. At that time, printed circuit techniques were appearing, and Saunders–Roe developed the idea of making a strain gauge by etching the grid from a thin foil of the appropriate resistance material. This foil strain gauge was found to have a number of distinct advantages, and was rapidly adopted, particularly by manufacturers and users in the USA. It opened the way for much more extensive industrial use of strain gauge techniques, and today it is by far the most widely used type of gauge worldwide.

Variations of the foil strain gauge are produced by die cutting, as opposed to etching, and by vacuum deposition techniques for particular purposes discussed later. The majority of development effort since the introduction of the foil gauge has, however, been toward improved control and understanding of gauge materials, characteristics and design.

BASIC OPERATING PRINCIPLE

All electrically conductive materials possess a strain sensitivity—defined as the ratio of relative electrical resistance change of the conductor to the relative change in its length—and therefore could be considered as possible strain gauge materials. The strain sensitivity is a function of the dimensional changes which take place when the conductor is stretched elastically (which will be essentially the same for different materials), plus any change in the basic resistivity of the material with strain.

The electrical resistance of a conductor is given by

$$R = \rho l / A \tag{1}$$

where R = resistance, l = length, A = cross-sectional area and ρ = resistivity.

Strain sensitivity (which, for a strain gauge, is defined as *gauge factor*) is a dimensionless relationship expressed mathematically as

$$F = \Delta R / R / \Delta l / l \tag{2}$$

where F = strain sensitivity, R = initial resistance, ΔR = change in resistance, l = initial length and Δl = change in length.

From these two formulae, the basic strain sensitivity can be established due to the dimensional changes, assuming that the resistivity (ρ) remains constant. If the conductor is stretched elastically, for a given change in length (Δl) there will be an associated reduction in cross-sectional area due to the Poisson effect. These two effects are additive in increasing the resistance of the conductor, and assuming a Poisson's ratio of 0·3 (which is approximately the same for most resistance materials) the sensitivity factor is about 1·6, i.e. $F = 1 + 2v$.

In fact, however, tests on various resistance materials have shown that they exhibit widely different values of strain sensitivity (Table 1). These

TABLE 1
STRAIN SENSITIVITY OF VARIOUS MATERIALS

Material	Trade name	Typical strain sensitivity
Copper–nickel (55–45)	Constantan Advance	+2·1
Nickel–chromium (80–20)	Nichrome V	+2·2
Nickel–chromium (75–20) plus iron and aluminium	Karma	+2·1
Iron–chromium–aluminium (70–20–10)	Armour D	+2·2
Nickel–chromium–iron–molybdenum (36–8–55·5–0·5)	Isoelastic	+3·5
Platinum–tungsten (92–8)	—	+4·0
Copper–nickel–manganese (84–4–12)	Manganin	+0·6
Nickel	—	−12·0
Iron	—	+4·0

variations indicate that the specific resistivity (ρ) of the materials must be affected by strain, or perhaps more directly by the associated internal stress in the material. Strain sensitivity is, therefore, a combination of the effects of geometric changes plus a resistivity change due to changing internal stresses.

Beyond the elastic limit of the material, however, the change in internal stress approaches zero, and Poisson's ratio (v) approaches 0·5. In this case, resistance changes due to strain are primarily due to dimensional changes and the strain sensitivity ($F = 1 + 2v$) approaches 2·0.

This means that materials which have a strain sensitivity appreciably different from 2·0 in the elastic range will have values approaching 2·0 in the plastic range, with the associated non-linearity which this variation implies. It would appear from this, therefore, that only those alloys which have a sensitivity of approximately 2·0 in the elastic range will remain essentially linear over a wide strain range, and this is generally true of the most commonly used strain gauge materials. Some materials which have an attractively high sensitivity in the elastic range are in fact highly non-linear, which means that the sensitivity varies with strain, rendering them undesirable for strain gauge purposes.

MATERIALS

The basic materials used in the manufacture of the resistance strain gauge are those of the grid (the strain sensitive resistance element) and the backing, which will be considered separately.

The Strain Sensitive Resistance Element
Some of the desirable features for a strain sensitive resistance element are:

(1) Linear strain sensitivity in the elastic range—for accuracy and repeatability.
(2) High resistivity—for smallest size.
(3) Low hysteresis—for repeatability and accuracy.
(4) High strain sensitivity—for maximum electrical output for a given strain.
(5) Low and controllable temperature coefficient of resistance—for good temperature compensation.
(6) Wide operating temperature range—for the widest range of applications.
(7) Good fatigue life—for dynamic measurements.

Table 1 summarises the strain sensitivity of a number of resistance materials which might be considered for strain gauge purposes. In practice, only a few alloys come close to meeting the most desirable features; and the main ones, with a summary of their characteristics, are as follows.

Copper–Nickel Alloy

The strain gauge alloy is generally referred to as constantan, although this name also applies to a slightly different composition, used for thermocouples. Other trade names include Advance, Cupron and Copel.

Of all modern strain gauge alloys, copper–nickel (55–45) is the oldest and still the most widely used. This reflects the fact that it has the best overall combination of properties needed for many strain gauge applications. This alloy has, for example, an adequately high strain sensitivity which is relatively independent of strain level and temperature. Its resistivity is high enough to achieve suitable resistance values in small grids, and its temperature coefficient of resistance is not excessive. In addition, it is characterised by a moderate fatigue life and a relatively high elongation capability. Very importantly, it can be self-temperature compensated to match a wide range of test material expansion coefficients.

Operating temperature range is generally between $-50\,°C$ and $+150\,°C$, but the alloy may exhibit continuous small changes in resistance with time, at temperatures in excess of $+70\,°C$. At higher temperatures it becomes progressively subject to oxidation, leading to significant resistance changes with time.

The strain range of copper–nickel in its standard form is up to $\pm 5\,\%$ and up to $20\,\%$ in a super-annealed form.

Nickel–Chrome Alloys

The nickel–chrome alloys are the next in popularity as strain gauge materials. Modified Karma or K-Alloy, which is a 75–20 nickel–chrome alloy with additions of iron and aluminium, is one of the most important of the strain gauge alloys. It has good fatigue life, significantly better than copper–nickel, excellent stability, and its overall performance characteristics are superior to any other alloy currently available.

Its temperature coefficient of resistance can be adjusted by heat treatment to provide temperature compensation on various materials over a wide strain range. Its strain sensitivity, which decreases with temperature, can be controlled within limits to compensate for the reduction in modulus of elasticity with temperature of materials used for precision transducer applications.

Its high resistivity enables smaller gauges to be made for a given resistance, or a higher resistance achieved for a given size, compared with copper–nickel.

The operating temperature range is between −269 °C to +290 °C for static measurements, and it can be used for short periods at temperatures as high as +400 °C.

Other forms of nickel–chrome are sometimes used at higher temperatures (above +400 °C), but because of progressive changes in resistance at these temperatures, their use is normally restricted to dynamic measurements where continuous resistance changes are less significant.

Platinum–8% Tungsten

Platinum–8% tungsten has been developed as a high temperature strain gauge alloy to overcome some of the problems with nickel–chrome. Unlike nickel–chrome, it does not undergo any metallurgical changes at temperatures up to +900 °C, so that its resistance remains essentially unchanged with time at high temperatures. It has a high temperature coefficient of resistance which cannot be adjusted, but it is repeatable. In addition, it has a higher strain sensitivity than copper–nickel or nickel–chrome but it is non-linear, so that strain range is limited generally to about ±0·3%.

Isoelastic Alloy

This is an alloy of nickel, iron, chromium and molybdenum which has the advantage of excellent fatigue life and relatively high strain sensitivity. It has, however, a high temperature coefficient of resistance and, associated with the high sensitivity, it is non-linear at strain levels above about 0·5%. Its main application is for dynamic measurements where its excellent fatigue life and high sensitivity contribute to better measurements under dynamic testing conditions.

Other Alloys

As indicated in Table 1, there are a number of materials and alloys which could be used for strain gauge purposes. Most of them are unsuitable for one reason or another, in that their overall characteristics do not match up to those which are available from the ones already described. A considerable amount of research, however, has been carried out on materials for use at high temperature. The majority of these have various limitations, and at the present time, platinum–8% tungsten appears to offer the best compromise of the known materials which have been evaluated.

Backing Material

The backing or carrier material of a strain gauge serves several very important functions, including:

(1) a means of handling the gauge and of firmly supporting the grid and the attachment tabs or lead wires;

(2) providing a readily bondable surface for bonding the gauge to the test material;

(3) providing electrical insulation between the grid and the test material;

(4) faithfully transmitting the strain from the test material via the adhesive, to the gauge grid.

These primary functions lead to the need for some very specific requirements of the properties of the backing to achieve optimum overall strain gauge performance, and these are:

(1) high shear modulus, with minimum thickness, to ensure complete strain transmission;

(2) strength and flexibility, to reduce the possibility of damage during installation, and to permit installation on curved surfaces;

(3) high elongation capability, for use at high strain levels into the plastic range;

(4) good bondability, for good 'peel' strength between the grid and backing, and good adhesion between the backing and test material;

(5) inherently high insulation resistance, to eliminate measurement problems and inaccuracies due to shunting effects of low insulation;

(6) good stability, with minimum creep;

(7) retention of all of the above characteristics over as wide a temperature range as possible.

Modern backing systems in various forms meet most of these requirements, and the following is a summary of the main materials which are or have been used as strain gauge backings.

Nitrocellulose–Paper

The first 'paper-backed' resistance strain gauges used this type of backing, which is the only standard backing that can be used with solvent-evaporation nitrocellulose adhesives. The performance of such backing/adhesive systems is, however, generally inferior to what is obtainable with modern resins, from which most present day strain gauge backings are made.

Epoxy Resins
Particularly for foil strain gauges, unfilled cast epoxy resins were the principal backing materials until about the mid 1960s, when they were largely replaced by polyimides. A specially pure epoxy, characterised by high rigidity and outstanding freedom from creep is, however, still used for strain gauges intended for high performance transducer applications. The somewhat brittle nature of this material, which limits the maximum strain to 2%, makes it more difficult to handle and unsuitable for general strain measurement purposes. Filled epoxy resins or, more particularly, combination epoxy–phenolics, are in wide use, and, for the best overall strain gauge performance over a wide temperature range, are reinforced with glass fibre. The presence of glass fibre adds to the strength and flexibility of the resin system. This type of backing can be used for static and dynamic strain measurements from $-269\,^\circ$C to $+290\,^\circ$C, and for short-term tests, up to $+400\,^\circ$C. Maximum strain capability is between 1% and 2%.

Phenolic Resins
Phenolic resin systems have been used for many years to produce fully encapsulated high performance strain gauges. From both performance and manufacturing points of view, however, they are now more normally compounded with epoxy resins as already discussed.

Polyimides
The most widely used general purpose backing available today is cast polyimide film. It is a stable, extremely tough and flexible material, and can be contoured to a very small radius. It can be used over a wide temperature range, typically $-195\,^\circ$C to $+175\,^\circ$C, and at elongations up to 20%, making it well suited to high strain measurements.

Removable Backing
The present practical maximum upper temperature limit of most organic resins useful for strain gauge backings is around $+400\,^\circ$C. For higher temperatures, temporary, removable or strippable backing is attached to the grid, but is used only for handling purposes. The temporary backing, which has a low peel strength, is removed from the grid prior to placement on an insulating layer of the appropriate cement. The material used for this temporary carrier is typically Teflon.

GAUGE CONSTRUCTION

The construction of an electrical resistance strain gauge involves bringing together the optimum electrical resistance material and backing in the best possible way from manufacturing, utilisation and performance points of view. The basic requirements for the gauge grid and backing materials have already been discussed, and it is appropriate to review the overall specifications for an ideal bonded resistance strain gauge. Some of the most important desirable features would be the following

(1) Small size and mass.
(2) Ease of production in different sizes and configurations.
(3) Robustness, with ease of handling and application.
(4) Good stability, repeatability and linearity over a wide strain range.
(5) Reasonable sensitivity to strain.
(6) Freedom from, or ability to control, effects of environmental variables such as temperature.
(7) Suitability for static and dynamic measurements and remote recording.
(8) Low cost.

Items (1), (2), (3) and (8) are features related to the construction of the gauge to be dealt with in this section. Items (4), (5), (6) and (7) are more specifically related to performance and will be dealt with later in this chapter and in other chapters.

From basic measurement circuit requirements, the resistance strain gauge element must have a resistance value which is not too low. For many years the accepted standard was 120 ohms. Today, however, more and more applications are with 350 ohms and many other higher values are available (typically 500, 600 and 1000 ohms). The relative significance of these higher values is discussed later.

Because of practical limitations on the diameter or cross-section of resistance elements, the total length needed to achieve the minimum desired resistance using a single straight filament would generally be considerably longer than the desired measurement length. Virtually every resistance strain gauge, therefore, consists of a grid configuration of a single wire. For a given required resistance and wire diameter, the overall grid length is a function of the number of grid lines. Since the grid will respond to the average value of strain between the ends of the grid, this dimension is the basic measuring length of a strain gauge, and is called the 'gauge length'.

Wire Strain Gauges

Flat-Grid Type

The 'paper-backed' wire strain gauge consists typically of a grid of resistance wire, wound on a special jig, to which the paper backing, coated with nitrocellulose cement, is attached (Fig. 2(a)). It is necessary, however, to support the delicate grid during handling and installation to prevent breakage of the fine wires and distortion of the grid. The grid is therefore normally sandwiched between two very thin layers of paper, and the assembly is completely impregnated with the nitrocellulose to give it strength and flexibility. In some cases, the top layer of paper is replaced by a felt pad about 1 mm thick. The bottom layer of paper also serves the very important purpose of electrically insulating the grid from the bonding surface.

General purpose wire gauges usually employ measuring grids of approximately 0·02 mm diameter copper–nickel wire, and use a nitrocellulose cement as the binder. Overall thickness of a gauge of this type is approximately 0·1 mm in the grid area.

Because it would be very difficult to attach instrumentation wires in the field to the fine diameter wire used in the grid, heavier wires of about 0·2 mm diameter are soldered to the grid for connection to the signal wires after gauge installation.

Many other types of wire gauges exist, involving different features to better meet special performance requirements. Replacing the paper–nitrocellulose backing with glass fibre and thermosetting resins, such as the epoxies or phenolics already discussed, will greatly extend the useful

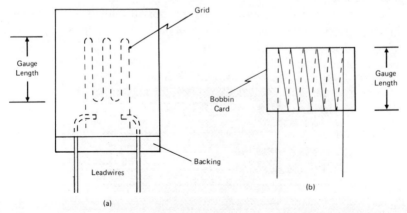

FIG. 2. Wire strain gauge outlines (a) flat-grid type, (b) wraparound grid type.

temperature range of the gauge. The grid may be of alloys other than copper–nickel for improved performance at very high or very low temperatures, or to exhibit better fatigue life under cyclic strain at high strain levels.

'Wraparound' Grid Type

An alternate form of wire strain gauge construction is also in limited use, generally in combination with the nitrocellulose–paper backing. This is called the 'wraparound' or bobbin type, in which the grid is wound on a porous bobbin card about 0·05 mm thick. Separate thin layers of paper are then adhered to both sides of the bobbin to serve as the backing and overlay of the completed gauge. Wraparound gauges are easier and cheaper to make, especially in smaller sizes, since the bobbin card controls the gauge length and supports the grid during fabrication. Performance of this type is inferior to the flat-grid construction, however, because the gauge is much thicker, and strain introduction into the end of each strand is poorer. Drying of the nitrocellulose adhesive is slower because of the greater thickness. Wraparound grid construction is illustrated in Fig. 2(b).

Foil Strain Gauges

The foil strain gauge (Fig. 3) is essentially a small printed circuit. Manufacturing techniques involve the preparation of artwork for the master gauge pattern to a very high degree of accuracy, many times larger than the finished gauge. This master is then reduced photographically to the correct size and reproduced as multiple negative images on a photographic plate.

Specially treated foil material of the appropriate alloy is rolled to exact thickness, typically 0·003 to 0·005 mm, and the backing material applied by

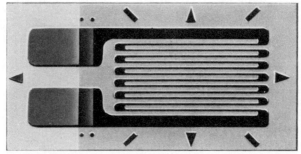

FIG. 3.   Modern foil strain gauge. Courtesy of Micro Measurements.

a casting or lamination process. A light-sensitive photographic emulsion is applied to the foil surface, and then exposed to the multiple image photographic plate, followed by a photographic developing process. The exposed portions of the foil corresponding to the grid of the gauge are now resistant to the subsequent rigidly controlled chemical etching process, which removes the alloy in the unexposed areas.

This leaves identical strain gauges on the sheet of backing material which are then separated for stringent electrical and optical inspection. Finishing work includes final resistance adjustment as necessary, and other steps, which may include encapsulation (with, for example, a film of the backing material), lead wire attachment, or other special processes.

The resulting gauge grid differs from that of a wire gauge in that the cross-sectional area is rectangular instead of circular. Most of the basic differences in performance between the foil and wire gauges derive from this difference as follows:

(1)   For the same cross-sectional area, the exposed surface area of the foil conductor is much greater, and correspondingly less unit shear stress is required in the backing and adhesive to strain the conductor. As a result, strain transmission into the grid is more complete for foil gauges.

(2)   A much better thermal path exists from the foil conductor to the substrate, and foil gauges can therefore operate at considerably higher power levels.

(3)   The width of the foil conductor is usually large compared to the thickness of the backing and adhesive layer, therefore, unlike wire gauges, transverse strains are transmitted to some extent into the 'active' part of the conductor.

One of the most significant advantages of foil gauge construction is in the manufacturing process itself. Wire gauges must still be manufactured largely by hand, and this places certain limitations on cost and on performance reproducibility from gauge to gauge. In contrast, the precision photoetching process used for foil gauge production permits many identical gauges of exact size and geometry to be formed at the same time from the sheet of specially treated foil, rolled to exact thickness. It also provides complete freedom in two dimensions for the design of single and multiple grid gauges of optimum geometry.

Figure 4 illustrates some typical examples of the wide variety of commercially available foil gauges. The shortest gauge length currently available is 0·008 in (0·20 mm) and the longest, 4·00 in (102 mm). Typical

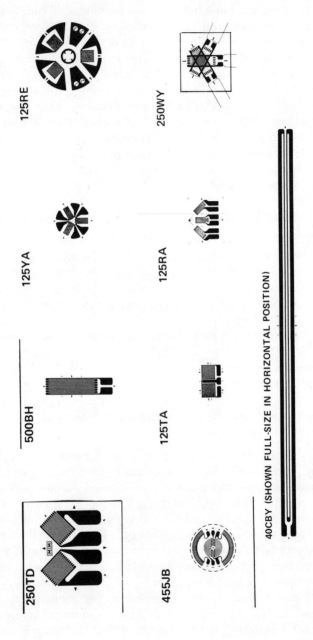

FIG. 4. Typical foil strain gauge patterns. Courtesy of Micro Measurements.

resistance values are 60, 120, 350, 500 and 1000 ohms, with some special purpose gauges at 1500 and 3000 ohms. Some recent developments in manufacturing techniques are now being used to develop very high resistance miniature strain gauges for the transducer market (e.g. sizes of 1·5 × 1·5 mm in resistances to 5000 ohms or more). Typical overall thicknesses for foil gauges are 0·03 to 0·07 mm, depending on the type of backing, encapsulation, etc.

Figure 5 illustrates another important advantage of foil gauges. The end loops may be extended considerably farther than the width of one strand. This reduces the resistance of the end loop, and makes the grid less sensitive to transverse strains. Also, and of perhaps greater importance, the electrically inactive portion of the large end loops represent a very large bonded area for introducing strain into the ends of the grid. As a result, very short gauge length patterns can be produced in foil construction without the serious lack of strain transmission which would be found in wire grids of the same size. Since the term 'gauge length' is intended to

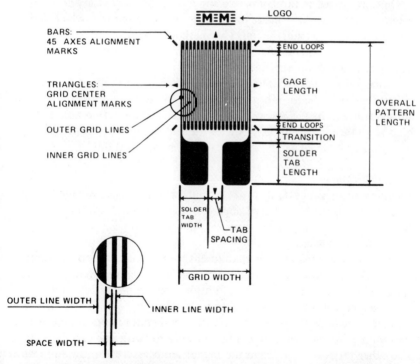

FIG. 5.   Gauge nomenclature and features of a typical foil strain gauge.

specify the length over which a grid measures strain, it is common practice to use the length inside the end loops of a foil gauge as defining the true gauge length (Fig. 5).

Two of the advantages that foil gauges normally have over wire gauges become disadvantages, however, in the special case of gauges designed for extremely high temperature operation. At temperatures high enough to require inorganic (ceramic) insulation and adhesives, it becomes very difficult to maintain insulation resistance between the grid and mounting surface at values high enough for proper operation. This problem is aggravated by the very large surface area of the foil conductors, which tends to increase electrical leakage to the specimen. Also, oxidation of the grid creates drift effects (resistance change with time) at these temperatures, and more surface area of a foil conductor is exposed to oxidation than is true of wire of comparable cross-sectional area. An additional practical consideration is that ceramic bonding agents are often more compatible in application and/or curing with the round cross-section of wire.

One alternative method of producing foil strain gauges is by die cutting, where an accurate die is made in the form of the gauge grid, which is then used to cut the grid from the foil. This method is mainly used for producing foil gauges from materials which are difficult to chemically etch, for example, platinum–tungsten.

Other foil gauge manufacturing methods include vacuum deposition, either directly in the form of the gauge grid, or as a very thin film which is subsequently etched. These processes are used primarily to produce small, high resistance gauges directly on to transducer elements, rather than as discrete bondable elements for stress analysis applications.

Weldable Gauges

Weldable strain gauges (Fig. 6) are special forms of gauges intended for use in difficult industrial and civil engineering environments where adhesive bonding of gauges is not practicable. They are also used for certain high temperature applications.

They are available in various different forms, according to the specific purpose for which they are intended. One is simply a standard foil type gauge, bonded under laboratory conditions to a stainless steel shim, which can then be spot-welded to the test structure 'on-site'. There are two versions, one of which is used for normal temperature applications where the strain gauge element is of copper–nickel alloy with polyimide backing. The other version is based on the higher temperature Karma alloy with glass fibre reinforced epoxy–phenolic backing. This is used for applications

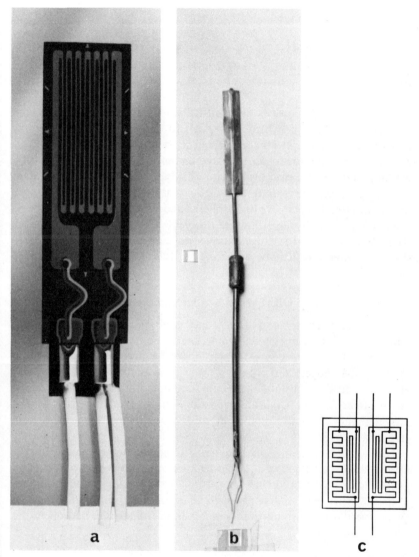

FIG. 6.   Weldable strain gauges (a) Micro Measurement, (b) Ailtech, (c) BLH.

at temperatures up to 260 °C, where otherwise it would be necessary with an ordinary gauge to apply elevated temperature curing cycles to conventional adhesives. In many large-scale structures this may be neither practical nor possible.

Another type of weldable gauge consists of a free-filament grid which has been bonded to a stainless steel shim under laboratory conditions, but this time with inorganic adhesives (e.g. ceramics) for use at temperatures in excess of +300 °C.

Yet another type consists of a small diameter stainless steel tube with flanges down each side which are welded to the test specimen. The tube contains a single loop wire strain element (typically platinum–tungsten), packed in magnesium oxide to provide both the strain transmission medium and the electrical insulation. This type of gauge can be used for operating temperatures up to 500 °C, but it is often specified also for use under water or in corrosive atmospheres and other hostile environments where temperature is not the major problem. This is possible because the construction permits the gauge to be hermetically sealed by means of steel-sheathed cable which can be swaged to the tube containing the resistance element.

PERFORMANCE CHARACTERISTICS

Performance characteristics of resistance strain gauges are functions of a number of different and generally unrelated factors. Not the least of these are the installation procedures (which depend on the expertise of the person applying the gauges) and adhesives, which are the subject of Chapter 2. However, current installation techniques and adhesives allow almost anyone to achieve a reliable and repeatable standard after a little practice, provided that the basic quality of the gauges themselves is of a high standard.

This aspect is completely in the hands of the manufacturer and can, in the first instance, be related to three basic measurable parameters which are—gauge factor, resistance value, temperature compensation, and tolerances of these. Other basic characteristics which generally form part of a strain gauge specification are maximum elongation, fatigue life, and operating temperature range.

Gauge Characteristics
Gauge Factor (and Transverse Sensitivity)
Gauge factor is largely governed by the basic strain sensitivity of the alloy from which the grid is manufactured, as already discussed. The actual

gauge factor is, however, more complex, and is a function of the metallurgical condition of the alloy, its temperature, the gauge grid size and configuration, and how it is measured.

In general, the gauge factor of a practical bonded strain gauge is lower than the unbonded strain sensitivity of the grid material. There are several reasons for this difference, one of them being the shear lag at the ends of the grid. It takes a finite distance from the ends of the foil to transmit the full strain in the test piece surface up through the thickness of the foil. The more lines there are in a gauge grid (and therefore more ends affected by the strain gradient) the lower the gauge factor will be, for a given gauge length or resistance. This effect can be reduced in foil gauges by suitable design. Another reduction in sensitivity is caused by the 'dead' resistance in the gauge. Only the active portion of the grid, i.e. the parallel lines, will change resistance with axial strain. The end loops do not respond to axial strain, but they still have a finite resistance. Hence the value of ΔR is reduced because R is the *total* gauge resistance.

The end loops in a wire or foil grid are also responsible for the characteristic referred to as 'transverse sensitivity'. This is the ratio of the gauge response perpendicular to the grid axis, with that parallel to the axis. It exists because each end loop represents a small amount of strain sensitive conductor at right-angles to the major grid axis. This portion of the grid will, therefore, be aligned with, and respond to, the lateral strain in the test surface. The transverse sensitivity due to this effect will normally be positive for a flat-grid wire type of strain gauge. 'Wraparound' wire grids usually have negative transverse sensitivities because the end loops are primarily along the axis perpendicular to the specimen surface.

Foil gauge grids normally have much less of this effect present because of the low resistance of the grid end loops. They are, however, affected by transverse strains to some extent, because the width of the grid line is often sufficient for lateral strain introduction through the backing. This lateral strain may produce either positive or negative resistance changes, depending on the alloy, and the magnitude will be a function of the grid line width. The overall transverse sensitivity of a foil gauge may, therefore, be positive or negative, depending on the relative contribution of the end loops, line width and the alloy. By proper balancing of these factors, foil gauges can be designed to have transverse sensitivities that are very low or effectively zero, but other considerations may make this less desirable.

The gauge factor of commercial bonded strain gauges is determined experimentally by installing several identical gauges from the same manufacturing batch on a test beam, to which is applied a known mechanical strain. The average ratio of resistance change to strain change is obtained to determine the gauge factor for that particular gauge type and batch. If a uniaxial strain field were used for this calibration, the value obtained would be slightly lower than for an equivalent straight length of the same alloy because of the 'dead' resistance component of the end loops as explained earlier. In normal practice, however, it is more usual and convenient to use a uniaxial stress field such that there is a lateral strain component in the test bar due to Poisson's ratio. This acts on the transverse sensitivity factor to modify slightly the measured gauge factor. By their very nature it is not possible to establish the gauge factor for every gauge, the figures being based on an average of samples from each particular manufacturing lot or batch, tested under tension or compression. The figures are usually quoted to two decimal places with a tolerance of $\pm 0.5 \%$.

Transverse sensitivity is usually measured with a special calibration rig designed to generate a pure uniaxial strain field. This figure is then quoted as the ratio between the transverse and longitudinal sensitivities. Some values of typical gauge factor and transverse sensitivity for commercial foil gauges are given in Table 2.

TABLE 2

EXAMPLES OF GAUGE FACTOR AND TRANSVERSE SENSITIVITY
VALUES. COURTESY OF MICRO MEASUREMENTS

Gauge type	Gauge factor	Transverse sensivity %
EA-06-250BG-120	2·11	+0·4
EA-50-250BG-120	2·125	+0·9
WK-06-250BG-350	2·05	−3·4
WK-06-500AF-350	2·04	−9·2
WK-15-125AD-350	2·16	−1·9

Gauge Factor Variation with Temperature

The alloys used in resistance strain gauges typically exhibit a change in gauge factor with temperature. In some cases the error due to this effect may be small enough to be ignored. In others, depending upon the alloy involved, the test temperature, and the required accuracy, correction for the variation may be necessary.

For example, the variation of copper–nickel alloy is essentially linear over its working temperature range, and the variation is approximately $+1 \%$ per 100 °C. In the case of Karma alloy, the variation is also almost linear with temperature, but the slope is negative and varies according to the self-temperature compensation characteristics of the material. This variation lies approximately between -0.7 and -1.3% per $100\,°C$. This feature of negative variation of gauge factor with temperature is used to produce 'modulus compensated' strain gauges for precision transducer applications. It is possible to make the percentage variation of gauge factor with temperature match the reduction of elastic modulus with temperature of some typical transducer body materials. This then has the effect of producing a constant transducer output with varying temperature for a given applied load.

Resistance

The resistance of the gauge is not in itself a performance characteristic, but the tolerances to which the manufacturer supplies the gauge from the standard value is a measure of the production quality control. Typical tolerance for an open-faced 120 ohm gauge of 6 mm gauge length is $\pm 0.15\%$. For encapsulated gauges, very small gauges, higher resistances and different alloys, the resistance tolerance may be increased to $\pm 0.3\%$ or higher.

Temperature Effects on Gauge Resistance

All electrical conductors have a temperature coefficient of resistance, which means that a strain gauge made from these materials will undergo a change in resistivity with temperature. In a bonded resistance strain gauge there is an additional effect producing a change of resistance with temperature. When a gauge is bonded to a test material which has an expansion coefficient different from that of the gauge material, a change in temperature will produce a strain in the gauge and hence, a resistance change, due to the difference in expansion. The combination of these changes of resistance which occur, is interpreted as a mechanical strain and is referred to as temperature induced 'apparent-strain'. This is one of the most serious potential sources of error in the practice of static strain measurements with strain gauges.

Some strain gauge alloys exhibit large 'apparent-strain' errors (Fig. 7) but others can be processed to produce very small resistance changes with temperature. Typically, both copper–nickel and modified nickel–chromium can be processed by heat treatment to adjust the basic temperature coefficient of resistance to compensate for the resistance change due to differential expansion effects on various materials. These are referred to as self-temperature compensated (STC) gauges.

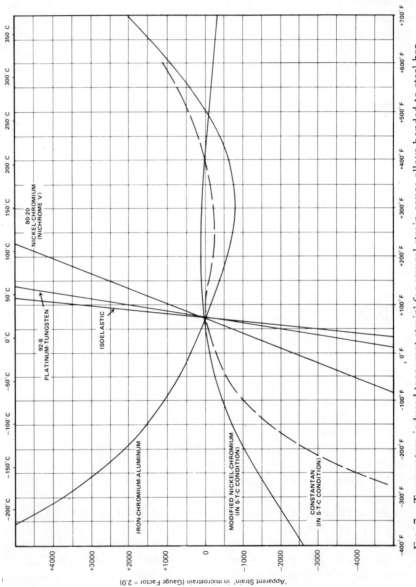

Fig. 7. Temperature induced 'apparent-strain' for several strain gauge alloys bonded to steel bar.

The manufacture of such gauges begins with selection of a melt of foil alloy that has all of the proper basic characteristics for accurate strain measurement, and, in addition, displays a certain specific temperature coefficient of resistivity in the rolled condition. This can be illustrated by using constantan alloy as an example. Figure 8 shows the behaviour of a particular melt of this alloy in both full-hard form, and in the fully annealed

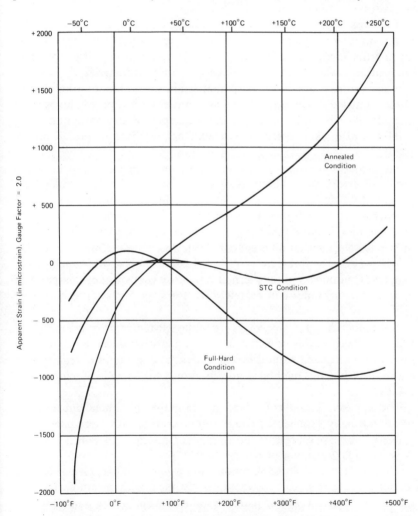

FIG. 8. Change in 'apparent-strain' characteristics of constantan alloy with heat treatment.

condition, mounted on a steel bar in both cases. The apparent-strain curve is fairly flat over the temperature range of about $-20\,°C$ to $+150\,°C$, and the slope of this portion of the curve is negative when full-hard, and positive when soft (fully annealed). By carefully selecting heat treatment cycles between these two extremes, the curve can be brought to an optimum position to show very low apparent-strain values when mounted on steel, as shown by the middle curve.

This type of processing is used to tailor batches of foil to show the best match on various materials with different expansion coefficients. Gauges made from these different batches of foil will then be self-temperature compensated when used on the intended specimen material.

A similar rotation of the apparent-strain curve can be produced in modified nickel–chromium alloys, though the thermal processing is more difficult since the alloy is much more complex than constantan. Very careful control, and special equipment, are required to obtain proper results.

The degree of rotation of the apparent-strain curve is generally chosen to minimise errors from about $-20\,°C$ to $+200\,°C$ for constantan STC gauges. It is evident from Fig. 7 that the useful region of self-temperature compensation extends over a considerably wider temperature span for the modified nickel–chromium alloy. In some cases, special test requirements may make a somewhat different curve rotation more useful in minimising errors over the temperature range of greatest interest. Mismatching the gauge STC number and the thermal expansion coefficient of the specimen material is sometimes deliberately done for such purposes as cryogenic tests.

Most modern foil strain gauges are of the self-temperature compensated type, produced in this way. The first STC gauge, however, consisted of a dual wire grid made from different alloys, having opposite temperature coefficients, but this method is now used only for some high temperature measurements.

The 'apparent-strain' curve (and generally the polynomial equation for it) are normally supplied by gauge manufacturers with every package of gauges. The curve represents the average for a number of gauges from the same manufacturing lot or batch.

There is, however, some variation from gauge to gauge, giving rise to a scatter band. The spread is approximately linear with deviation from the reference temperature. For gauges produced from constantan alloy, the scatter is approximately ± 0.27 microstrain per $°C$ and for Karma gauges ± 0.45 microstrain per $°C$, over the temperature range $0\,°C$ to $+175\,°C$.

Elongation Capability

All bonded resistance strain gauges are limited in the maximum strain level that can be applied before failure—by characteristics of the grid, the alloy, the backing and the adhesive. High performance gauges with glass fibre reinforced backings are usually limited to maximum strains of 1 or 2 %. General purpose self-temperature compensated gauges with copper–nickel grids are normally capable of measuring strains of up to 5 % in other than very small grid sizes. When the alloy has been super-annealed, however, the measurement of strains up to 20 % is possible, more particularly on metals and other high modulus homogeneous materials.

Weldable gauges, and high temperature gauges, are generally limited to strains of about 0·5 %, due to the nature of their construction or to the particular adhesives that have to be used for high temperature service.

Fatigue Life

In common with all metals, strain gauge grids will eventually fail in fatigue due to cyclic straining at reasonably high strain levels. Wire gauges become completely open-circuited within a few cycles of attaining their fatigue limit. Foil gauges usually remain intact for hundreds or thousands of cycles, however, before complete grid failure occurs because cracks at the line edges progress gradually across the width. This first manifests itself as an irreversible increase in resistance, producing a positive zero shift, which increases with additional cycling.

The presence of these cracks in the foil grid also produces an apparent increase in sensitivity. This is because the opening and closing of the cracks produces change in resistance which is not proportional to the normal resistance changes due to strain. This phenomenon is referred to as 'supersensitivity'.

Because of these considerations, the fatigue limit for foil strain gauges is normally defined as that point at which the zero shift has a value of between 100 and 300 microstrain, depending on the required test accuracy.

Fatigue life depends not only on the gauge materials and strain level, but also on gauge size, lead wire attachment, and uniformity of strain over the gauge area. Some typical strain gauge fatigue life data is given in Fig. 9.

Operating Temperature Range

The operating temperature range is a function of the gauge alloy, backing and adhesive, and a definition of other performance limitations associated with the temperature, such as stability. Typical temperature ranges for

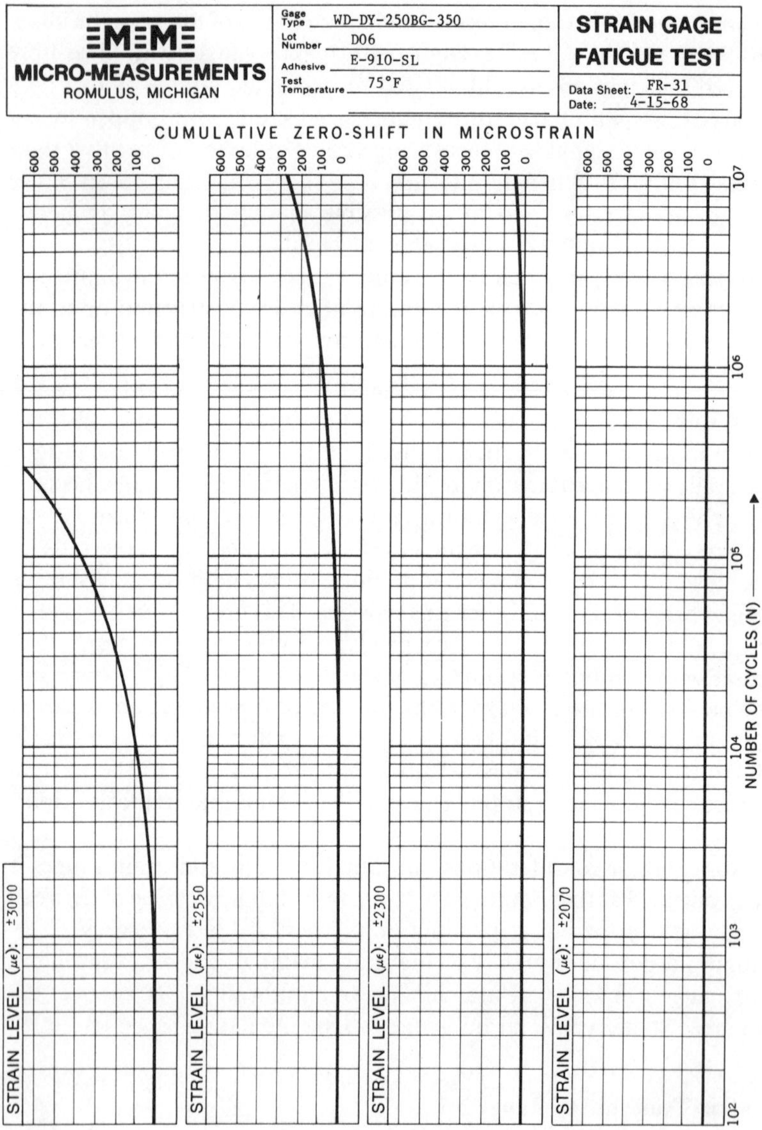

FIG. 9.   Fatigue life of a 6 mm gauge (isoelastic alloy with reinforced expoxy–phenolic backing).

TABLE 3

Foil	Backing	Static	Dynamic
Copper–nickel	Polyimide	-75 to $+175\,°C$	-195 to $+205\,°C$
Copper–nickel	Epoxy	-45 to $+99\,°C$	-195 to $+175\,°C$
Isoelastic	Polyimide	—	-195 to $+205\,°C$
Karma	Glass fibre reinforced	-269 to $+290\,°C$	-269 to $+400\,°C$
Nickel–chromium	Strippable	-269 to $+425\,°C$	

various foil/backing combinations of commercially available foil strain gauges are given in Table 3.

However, the upper temperature limits depend on many factors including the time at temperature. The upper limits quoted by many manufacturers will give a life measured in minutes rather than hours or days. If good stability is required for a period of months, the temperature limit should certainly be below 300 °C but what is meant by good stability? To some it may mean tens of microstrain to others hundreds, to some ±5 microstrain in which case the limit will be below 200 °C and perhaps below 100 °C depending on other factors. For purely dynamic strains there have been successful measurements at over 900 °C for short periods.

GENERAL STRAIN GAUGE OPERATING CHARACTERISTICS

Apart from those characteristics already outlined, which form part of a specification for a strain gauge, there are other main groups of behaviour to consider. One group relates to those performance parameters which are inherent in the gauge/adhesive combination, including linearity, hysteresis, zero shift, drift, creep, stability and resolution. Another group relates to the effects of environmental conditions which are dealt with separately.

Linearity, Hysteresis and Zero Shift
These three factors, which are generally interrelated, are functions of the characteristics of the strain sensitive alloy, the backing material, strain level, and adhesive cure. For experimental stress analysis, modern high performance strain gauges with well-cured adhesives can generally be considered free from these effects.

For properly installed gauges, deviations from linearity will generally be between 0·05 and 0·1 % of the maximum strain, and hysteresis and zero shift less than 0·2 % of the maximum strain.

In high performance transducers, however, small deviations of this order are significant, and can be minimised by post-curing the adhesive at a temperature higher than the maximum operating conditions, and by load cycling over the working range. The design and material of the transducer element itself are also very important factors.

Drift, Creep and Stability

Drift is a time-dependent and generally irreversible change in gauge resistance with time at constant temperature under no-load conditions (i.e. with zero applied strain). Properly manufactured modern foil gauges are capable of showing resistance drift of only a few parts per million per year. All strain gauge alloys, however, display increased drift rates as temperatures are increased.

Creep differs from drift in that it is a change of resistance with time at constant temperature, but under load (i.e. with applied strain), and is a function of the behaviour of the organic backing and adhesive materials. It can generally be minimised by post-curing the gauge installation above its maximum operating temperature, and is frequently less than the creep in the material being tested. Adjustments can, however, be made in foil gauge designs to compensate for material creep behaviour in precision load cells.

Stability can be defined as the lack of drift or creep over long periods of time, typically months or years. This is particularly important when it is necessary to record strains over long periods without the opportunity to unload the test specimen or structure and check zero references. This is one of the more difficult applications of strain gauges, but with proper care and attention excellent long-term stability can be obtained with modern strain gauges, adhesives, and protective coatings. Experiments have shown stability to within thirty parts per million (equivalent to a strain of 0·003 %) over 900 days for Karma alloy with glass fibre reinforced epoxy–phenolic backing and a well cured epoxy adhesive system.

Resolution

Resolution, or the ability of the gauge to detect small changes in strain, could be considered as being infinite. However, limitations of instrumentation and other performance factors generally mean that a resolution of 0·1 microstrain (one part in ten million) is about the smallest practical value attainable.

Heat Dissipation

By its very nature, it is necessary to pass an electrical current through a resistance strain gauge to provide an output which, in turn, is proportional to the gauge excitation level. For maximum sensitivity and resolution, the highest practical excitation level is desirable, although the resulting current in the gauge can produce undesirable self-heating effects. This is a function of the gauge size, resistance, and heat sink capability of the test specimen.

Under normal conditions, essentially all of the heat developed in the gauge flows through the backing and adhesive into the specimen. Metallic specimens are fair-to-excellent heat sinks, depending on their size, shape, and thermal conductivity. Unfilled plastics, on the other hand, are thermal insulators rather than heat sinks, so that much of the heat from the gauge must be transferred to the surroundings by convection.

Errors due to self-heating effects show up in several ways, but primarily as instability in readings of the gauge output. Voids in the adhesive glue line, or imperfections in the gauge grid, will result in local 'hot spots' with consequent reductions in performance. In the case of plastic specimens, the self-heating of the gauge installation may cause a severe local loss in specimen properties (e.g. elastic modulus), leading to inaccurate and misleading results.

The ability of the gauge to dissipate heat for given heat sink conditions is a function of the area of the gauge grid. It is usual to recommend, therefore, approximate levels of power density for various specimen and test conditions, and convert these to the corresponding excitation voltages for a given gauge size and resistance. Table 4 lists typical power density recommendations for foil strain gauges. A power density of 0·01 watts/ mm^2 will produce a temperature increase of about 1·5 °C for a typical foil gauge mounted on a good heat sink.

Different power density requirements apply, depending on the degree of accuracy required from the measurements and whether they are static or dynamic

In particular, when purely dynamic signals are involved, minor instability and creep are relatively less important, so that considerably higher power levels can be employed. This can be especially important in helping to maximise the signal-to-noise ratio in dynamic measurements where other extraneous influences can produce significant unwanted noise levels.

Most present day strain gauge instrumentation provides either fixed low level excitation voltage, or variable voltage, starting at a low level, in order to obtain the best compromise between self-heating and signal level.

TABLE 4
ALLOWABLE GAUGE POWER DISSIPATION IN WATTS/IN.² (KILOWATTS/m²). COURTESY OF MICRO MEASUREMENTS

Accuracy	Heat sink, watts/in.² (kilowatts/m²)				
	Excellent (heavy aluminium or copper)	Good (thick steel)	Fair (thin stainless steel or titanium)	Poor (filled plastic such as fibreglass/epoxy)	Very poor (unfilled plastic such as acrylic or polystyrene)
Static					
High	2–5 (3·1–7·8)	1–2 (1·6–3·1)	0·5–1 (0·78–1·6)	0·1–0·2 (0·16–0·31)	0·01–0·02 (0·016–0·031)
Moderate	5–10 (7·8–16)	2–5 (3·1–7·8)	1–2 (1·6–3·1)	0·2–0·5 (0·31–0·78)	0·02–0·05 (0·031–0·078)
Low	10–20 (16–31)	5–10 (7·8–16)	2–5 (3·1–7·8)	0·5–1 (0·78–1·6)	0·05–0·1 (0·078–0·16)
Dynamic					
High	5–10 (7·8–16)	5–10 (7·8–16)	2–5 (3·1–7·8)	0·5–1 (0·78–1·6)	0·01–0·05 (0·016–0·078)
Moderate	10–20 (16–31)	10–20 (16–31)	5–10 (7·8–16)	1–2 (1·6–3·1)	0·05–0·2 (0·078–0·31)
Low	20–50 (31–78)	20–50 (31–78)	10–20 (16–31)	2–5 (3·1–7·8)	0·2–0·5 (0·31–0·78)

The power density is expressed here in terms of kilowatts per square metre, in accordance with recommended SI practice. It is numerically equal to the power density in milliwatts per square millimetre.

$$\text{Power Density} = P'_G = \frac{E_B^2}{4R_G A_G}$$

where: E_B = bridge excitation voltage; R_G = gauge resistance; and A_G = grid area [(active gauge length) × (grid width)].

Note: power density recommendations for specific gauge types are given in *Strain Gauge Excitation Levels*, TN-127, Micro Measurements Division, Measurements Group Inc., Raleigh, USA.

Reinforcement

The reinforcing effect of a strain gauge on most metallic specimens can be considered negligible. The composite modulus of elasticity of the grid and backing is generally about one-tenth that of most structural metals, and the thickness is minimal. In those cases where the specimen has a low modulus and/or is very thin, however, the operating force required by the gauge may significantly alter the strain distribution around it. This reinforcement effect can, therefore, be a problem with plastic specimens or on bending-beam type low force strain gauge transducers.

In the case of a transducer element, reinforcement effects can be taken care of in the calibration. In other cases it is desirable, if possible, to check for the effect on a test specimen in a testing machine.

Hydrostatic Pressure

In those cases where strain gauges are employed in pressure vessels or pipelines exposed to gas or fluid pressure, resistance changes can occur due to these effects alone. Pressure effects have been treated by various experimenters, and although different figures have been obtained, they have been shown to be generally small. A typical figure for a copper–nickel gauge is about -100 microstrain per kilobar.

Where very high pressures are involved, data needs to be corrected for these effects, but the most important single consideration is the elimination of voids or bubbles in the adhesive layer or gauge backing. Under high pressures, these will cause local deformations which will produce non-linear outputs and poor zero returns, and even result in gauge failure. It is essential, therefore, that particular care is taken in every aspect of the selection and application of strain gauges for measurements in high performance situations.

Thermal EMFs

It is possible for a thermocouple junction to be formed by the presence of dissimilar metals where strain gauge lead wires are attached to the gauge tabs. When copper lead wires are used with a copper–nickel gauge, for instance, a typical thermocouple combination arises, such that changes in temperature can produce small d.c. voltages.

Normally these effects tend to cancel out because of the two opposing lead wire connections on a standard strain gauge, provided that both connections are at the same temperature. With modern strain gauges and protective coatings where this is generally the case, the effects can be ignored.

Where there are special problems of temperature variations and gradients, the use of a gauge made from a different alloy (e.g. Karma) can significantly reduce the effect. In addition, the use of an a.c. carrier circuit will eliminate measurement of spurious d.c. signals due to this effect, or instrument circuits can be designed to enable the presence of them to be detected. For example, an on/off switch for the bridge power allows the presence of 'noise' to be detected and measured.

Magnetic Fields
Errors can be introduced into the strain gauge measurements by strong magnetic fields. There are various ways of overcoming this problem by paying particular attention to the gauge wiring and shielding. Use of strain gauges made from Isoelastic alloy, which is a magnetic material, should be avoided however.

A special approach to this problem exists in the form of a special strain gauge consisting of two identical grids superimposed on top of each other, separated by an insulating layer, and then connected in series. This arrangement results in almost complete cancellation of voltages which may be induced in the gauge due to magnetic fields.

STRAIN GAUGE SELECTION PARAMETERS

The installation and operating characteristics of a strain gauge are influenced by all of the various parameters which have already been reviewed. The process of selecting a gauge for a particular measurement situation consists of determining the compromise of the combination of parameters which are compatible with environmental and other operating considerations allied to the installation and operating requirements, such as accuracy, stability, maximum elongation, test duration, cyclic endurance and simplicity and ease of installation.

The question of the most suitable alloy, backing material and adhesive can generally be determined from manufacturers' selection tables. This still leaves, however, a wide choice in gauge length, pattern and resistance, and Table 5 lists the influence of size, resistance, and end loop size on some of the main performance criteria which have already been discussed. Some of the more detailed considerations are as follows.

Gauge Length
Gauge length is an important consideration in strain gauge selection, and often the first parameter to be defined. Strain measurements are usually

TABLE 5

EFFECTS OF GAUGE SIZE, RESISTANCE AND END LOOP LENGTH ON FOIL GAUGE PERFORMANCE CRITERIA. COURTESY OF MICRO|MEASUREMENTS

Performance criterion	Desirable for optimum performance		
	Gauge size	Gauge resistance	End loop length
Gauge factor			
(a) uniformity	Large	Medium	Medium
(b) maximum	Medium	High	Long
Low transverse sensitivity	Small	High	Long
Apparent strain uniformity	Medium	Medium	Medium
High strain	Medium	Low	Short
High fatigue life	Large	Low	Short
Low creep	Medium	Medium	Medium
Maximum temperature limit	Large	High	Long
Low self-heating	Large	High	Medium
Ease of installation	Medium	Low	—
Low 'noise'	—	High	—
Low cost	Medium	Low	—

made at the most critical points on a machine part or structure—that is, at the most highly stressed points. Very commonly, the highly stressed points are associated with stress concentrations, where the strain gradient is quite steep, and the area of maximum strain restricted to a very small region. Since the strain gauge tends to integrate, or average, the strain over the area covered by the grid, and since the average of any non-uniform strain distribution is always less than the maximum, a strain gauge which is noticeably larger than the maximum strain region will indicate a strain magnitude which is too low.

Larger gauges, however, when they can be employed, offer several advantages. To begin with, they are usually very much easier to handle (in gauge lengths up to, say, 13 mm) in nearly every aspect of the installation and wiring procedure than miniature gauges. Furthermore, large gauges provide improved heat dissipation because they introduce, for the same nominal gauge resistance, lower wattage per unit of grid area. This consideration can be very important when the gauge is installed on plastic or other material with poor heat transfer properties, as previously discussed.

A special application of very large gauges (up to 100 mm) is in strain measurement on non-homogeneous materials. When measuring strains in a concrete structure it is ordinarily desirable to use a strain gauge of sufficient gauge length to span several pieces of aggregate in order to measure the representative strain in the structure. It is usually the average strain that is sought in such instances, not the local fluctuations in strain occurring at the interfaces between the aggregate particles and the cement. In general, when measuring strains on structures made of composite materials of any kind, it is preferable to select a gauge length which is large with respect to the dimensions of the inhomogeneities in the material.

Gauge Pattern
This includes the shape of the grid, the number and orientation of the grids in multiple grid gauges, the solder tab configuration and arrangement, as well as various construction features which are standard for a particular gauge pattern.

In experimental stress analysis, a single grid gauge would normally be used only when the stress state at the point of measurement is known to be uniaxial and the directions of the principal axes are known with reasonable accuracy ($\pm 5°$).

For a biaxial stress state—a common case necessitating strain measurement—a two or three element rosette is required in order to determine the principal stresses. When the directions of the principal axes are known in advance, a two element 90° (or 'tee') rosette can be employed, with the gauge axes aligned to coincide with the principal axes. The directions of the principal axes can sometimes be determined with sufficient accuracy from one of several considerations. For example, the shape of the test object and the mode of loading may be such that the directions of the principal axes are obvious from the symmetry of the situation, as in a cylindrical pressure vessel. The principal axes can also be defined by testing with a brittle lacquer or photoelastic coating.

In the most general case of surface stresses, when the directions of the principal axes are not known, a three element rosette must be used to obtain directions with magnitudes of the principal stresses. The rosette can be installed with any orientation, but is usually mounted so that one of the grids is aligned with some significant axis of the test object. Three element rosettes are available in both 45° rectangular and 60° delta configurations. The usual choice is the rectangular rosette since the data reduction is somewhat simpler for this configuration. Many special purpose strain gauge designs are simply intended to ease the installation process to permit two or more grids to be installed at the same time with precise spacing and alignment. These include gauges used for measuring torsion in shafts, multi-grid strip gauges for use in regions of high strain gradients, and gauges for transducers. Special designs to fit particular strain field geometries include diaphragm gauges for use on diaphragm pressure transducers.

Gauge Resistance

In certain instances, the only difference between two gauge patterns available in the same series is the grid resistance—typically 120 ohms versus 350 ohms. When the choice exists, the higher resistance gauge is preferable in reducing the heat generation rate by a factor of three (for the same applied voltage across the gauge). Higher gauge resistance is also advantageous in decreasing lead wire effects such as circuit desensitisation due to parasitic resistance in the lead wires, and spurious signals caused by lead wire resistance changes with temperature fluctuations. Similarly, the signal-to-noise ratio can be improved with high resistance gauges (by using higher voltages) when the gauge circuit includes switches, sliprings, or other sources of random resistance change.

OTHER SPECIAL TYPES OF RESISTANCE STRAIN GAUGE

The following is a brief review of some of the special types of resistance strain gauge which have been developed to meet particular measurement problems.

Self-adhesive Gauges

This type of gauge consists of a vacuum deposited gold film embedded in an elastomer. It works on the principle of self-adhesion between a plate polished elastomer and another polished surface of harder material. The bond is of appreciable shear strength, so that it can readily and repeatedly be removed and replaced. Because of a high temperature sensitivity, the self-adhesive gauge is restricted to applications such as the exploration of strain on a low frequency cyclically loaded specimen.

Flexagage

The flexagage consists of two identical strain gauges mounted exactly opposite on either side of a strip of plastic material of accurately known thickness. When bonded to a structure, any bending present will result in a difference in the strains measured by the two gauges. It is intended for the investigation of strains in skins and shells where access is possible from only one side, and separates bending from plane strains.

Fatigue Life Gauge

The S/N® Fatigue Life Gauge has the general appearance of a standard foil gauge and is manufactured with the same basic processes. The foil material is specially processed such that when bonded to a structure and subjected to cyclic strains, it undergoes irreversible resistance changes which are a function of strain levels and number of cycles. These permanent resistance changes, which can be monitored periodically without the need for continuous connections to any instrumentation, provide information related to the load life history of the structure.

Manganin Gauges

Manganin is an alloy of copper–nickel–manganese which, as a conventional strain gauge material, has lower sensitivity than copper–nickel. Its resistivity is, however, a linear and repeatable function of applied pressure (approximately 0·27 % per kilobar). Manufactured in the form of a strain gauge grid, it is used primarily for monitoring pressure waves normal to its surface arising from high level explosive blasts.

CONCLUSIONS

The modern foil type of electrical resistance strain gauge, whilst outwardly simple in principle and appearance, is a highly sophisticated measurement tool. What is particularly noteable is that in a time span of more than 40 years, which has seen so much technological development in general, the simple relationship between strain and resistance of an electrical conductor has remained at the forefront of measurement technology.

That it has done so is due to much patient work in studying and controlling the mechanical and metallurgical characteristics of resistance alloys and the behaviour of polymers. This has led to remarkable developments in the accuracy of resistance strain gauge based load cells and weighing systems.

With proper use and understanding, the strain gauge is capable of almost limitless application to any engineering problem involving the measurement of stress, strain, load, force, pressure, torque and deflection. In fact, probably the only limit to its utilisation is that of the imagination.

BIBLIOGRAPHY

General

DALLY, J. W. and RILEY, W. F., *Experimental Stress Analysis*, 2nd edn, Chapter 6, McGraw-Hill, New York, 1978.
HEARN, E. J., *Strain Gauges*, Merrow Publishing, Watford, UK 1971.
NEUBERT, H. K. P., *Strain Gauges, Kinds and Uses*, Macmillan, London, 1967.
PERRY, C. C. and LISSNER, H. R., *The Strain Gage Primer*, 2nd edn, McGraw-Hill, New York, 1962.
POPLE, J., *BSSM Strain Measurement Reference Book*, British Society for Strain Measurement, Newcastle, UK, 1979.
PROCTER, E., *Strain Gauge Practice, Methods and Practice for Stress and Strain Measurement. Part 1. Measurement of strain load and temperature*, British Society for Strain Measurement (Monograph), Newcastle, UK, April 1977.
STEIN, P. K., How to select a strain gage, *Strain Gage Readings*, 2 (No. 1), 1959.
Strain Gage Selection, 1976, TN-132-2, Micro Measurements Division, Measurements Group Inc., Raleigh, USA.
WEYMOUTH, L. J., STARR, J. E. and DORSEY, J., *Bonded Resistance Strain Gages, Manual on Experimental Stress Analysis*, 3rd edn, Chapter 2, Society for Experimental Stress Analysis, Westport, CT, USA, 1978.
WINDOW, A. L., *Developments in Stress Analysis—$l_{p|}$* Chapter 4, Applied Science Publishers, London, 1979.

Materials

BERTODO, R. R., Resistance strain gauges for the measurement of steady strains at high temperatures, *Proc. Inst. Mech. Eng.*, 1965, **178** Part 1 (No. 34).
EASTERLING, K. E., High temperature resistance strain gauges, *Brit. J. Appl. Phys.*, **4**, 1963, 79.
GRINDROD, A. and NODEN, J. D., Resistance materials for high temperature strain measurement, *Strain*, (1968) **4** (No. 3).
PITTS, J. W. and MOORE, D. G., Development of High Temperature Strain Gages, *NBS Monograph 26*, 1961.
STEIN, P. K., Constantan as a strain-sensitive material, *Strain Gage Readings*, **1** (No. 1), 1958.
STEIN, P. K., Iso-elastic as a strain-sensitive material, *Strain Gage Readings*, **1** (No. 3), 1958.

Characteristics

Errors Due to Transverse Sensivitiy Errors in Strain Gages, 1980, TN-137-2, Micro Measurements Division, Measurements Group Inc., Raleigh, USA.
FAIRBAIRN, J., Creep strain measurements by foil gauges on aluminum alloys at 150 °C to 210 °C, *Strain*, 1974, **10** (No. 4).
Fatigue of Strain Gages, 1974, TN-130-3, Micro Measurements Division, Measurements Group Inc., Raleigh, USA.
FREYNIK, H. S. and DITTBENNER, G. R., Strain gage stability measurements for years at 75 °C in air, *Experimental Mechanics*, April 1976.

MARSCHALL, C. W. and HELD, P. R., Measurement of long-term dimensional stability with electrical resistance strain gauges, *Strain*, 1977, **13** (No. 1).

MEYER, M. L., A simple estimate for the effect of cross-sensitivity on evaluated strain gauge measurements, *Experimental Mechanics*, November 1967, 476–80.

MILLIGAN, R. V., The effects of high pressure on foil strain gauges, *Experimental Mechanics*, 1964 **4** (No. 2), 25.

MILLIGAN, R. V., The effects of high pressure on foil strain gauges on convex and concave surfaces, *Experimental Mechanics*, 1965, **5** (No. 2).

MITCHELL, D. H., Strain and temperature measurements on epoxy resin models, *Transducer Technology*, 1979, **1** (No. 7).

NICKOLA, W. E., Strain gauge measurements on plastic models, BSSM Conference, September 1978, University of Bradford.

Optimising Strain Gage Excitation Levels, 1979, TN-127-4, Micro Measurements Division, Measurements Group Inc., Raleigh, USA.

Strain Gage Temperature Effects, 1976, TN-128-3, Micro Measurements Division, Measurements Group Inc., Raleigh, USA.

WU, C. T., Transverse sensitivity of bonded strain gages. *Experimental Mechanics*, 1962, **2** (No. 11), 338–44.

Chapter 2

ADHESIVES AND INSTALLATION TECHNIQUES

G. C. MORDAN

Welwyn Strain Measurement Ltd, Basingstoke, UK

INTRODUCTION

The modern foil strain gauge is probably one of the most useful and widely used investigative tools available to engineers today.

At first sight it may appear a simple device but when all its characteristics are fully understood together with the effects of the operating environment it can be successfully used for a wide variety of applications including many involving hostile conditions.

However, no matter how well the strain gauge is understood its eventual performance will be critically dependent on the quality of the installation as the gauge can only perform as well as the installation will permit.

Selection of the correct application techniques and associated materials are therefore of great importance to the success of any strain gauge test programme.

With the availability of modern materials and techniques applying a strain gauge is no longer a black art and can be confidently undertaken by the first time user provided there is an understanding of the problems involved.

The basic process of applying a gauge to the test surface can be conveniently divided into the following major steps

(1) planning;
(2) selection of installation material;
(3) preparation of test surface;
(4) gauge preparation;
(5) application of adhesive;

39

(6)   clamping;
(7)   curing and post curing;
(8)   lead wire attachment and soldering;
(9)   application of protective coating system;
(10)   verification of installation prior to service.

Each step can be considered as a link in a chain and a weakness in any one step may eventually lead to failure of the installation in service.

The application techniques described in the following chapter have been developed for general purpose stress analysis and are easy to learn and capable of being used in almost any situation while at the same time producing consistent and reliable bonds and minimising errors that can be caused by poor techniques.

Having once learnt the methods described, there is much to recommend their adoption as a standard technique to be used by all technicians within an organisation.

Whilst this approach requires an investment in time to train personnel and maintain the level of training, it greatly assists in reducing 'technician factor' and contributes to the maintenance of consistent standards which in turn leads to an increase in confidence in the quality and reliability of every gauge installed.

It is also important to include personnel responsible for specifying test programmes so that they have an appreciation of the problems faced by those responsible for the practical work.

Finally it is possible, particularly with site work, that an occasion will arise when circumstances dictate some deviation from a recommended technique. In these cases it is essential before making any changes to be able to recognise and understand the possible effects that any short cut or compromise may have on overall performance.

PLANNING

The primary object of planning is to establish a knowledge of all the aspects of an application that might affect the choice of installation materials or techniques to be used. The difficulty is getting all the information needed.

Table 1 indicates the principal information required and the main decision areas each item may influence. For many tests the information required will be easily obtained and correct selection of material and techniques should present no problems.

However, there are occasions when accurate information about one or more important parameters is not available, and it may not always be possible to establish them properly before the test is started. In these cases there is no option open other than to assume worst case conditions and to make it clear to those conducting the test what the limitations of the installation are.

TABLE 1

Information	*Area of influence*
Type of test material	Selection of correct gauge STC number
	Surface preparation method
	Degreasing method
Nature of test surface, i.e. porous,	Surface preparation method
non-homogeneous poor heat sink,	Degreasing method, type of adhesive,
etc.	length, area and resistance of gauge
Operating temperature	Gauge and all installation materials
	Installation technique
Maximum strain level	Surface preparation method, choice of
	adhesive, protective coating and lead
	wire system. Type of gauge
Duration of test	Selection of adhesive and protective
	coating. Possibly gauge choice
Test environment	Selection of adhesive, protective coating
	and lead wires
Strain gradient at gauge location,	Length of gauge
Number of test cycles	Choice of gauge
Nature of environment in which	Selection of special gauge options
installation is to be carried out	Adhesive and protective coating
Time schedule allowed for	Selection of special gauge options
application of gauges	Adhesive and protective coating

If there is a possibility of changes in the initial test specification occurring during or after the test, such as increasing the duration or widening of the temperature range, these should be taken into consideration during the planning stage.

Finally record sheets should be prepared to allow details of the installation to be properly recorded and kept for reference. This should include information such as batch or lot number of strain gauges, adhesives and protective coatings used along with details of cure schedules, installed resistance values and resistance-to-ground.

ADHESIVE SELECTION

It is difficult to over emphasise the importance of the adhesive used to install a strain gauge as faithful transmission of the strain in the test surface through the adhesive layer into the gauge grid is of paramount importance.

Resistance strain gauge theory assumes that the strain appearing in the gauge grid is identical to that occurring on the surface to which the gauge is bonded. If this is not the case it will be impossible to produce accurate results from the data produced no matter how it is processed.[1] It is also important to recognise the significant influence the adhesive has on many installation properties, such as gauge factor, creep performance, linearity, hysteresis, zero shift with temperature, response to transient temperatures, installation resistance and heat dissipation characteristics. Manufacturers' mixing instructions should also be carefully observed as incorrect resin and hardener ratios can alter the physical properties of the cured adhesive. Therefore proper adhesive selection and careful application are one of the most important steps in the installation process. Although there are many adhesives available for both industrial and domestic use very few possess the necessary characteristics that make them suitable for installing strain gauges and their use normally results in an inferior performance. For this reason gauge manufacturers prefer to formulate their own adhesives with the correct characteristics over which they can exercise rigorous quality control to ensure consistent performance. In the few cases where a commercially available adhesive does perform satisfactorily gauge manufacturers normally test batches and only release for use those batches that meet the standards required. The cost of using the proper adhesive is small compared with the total cost of the test programme and compared with the risk of invalidating the data collected.

The ideal strain gauge adhesive would possess the following characteristics:

(1) |Be capable of forming a thin, void free glue-line with high shear strength.
(2) Be compatible with all gauge backings, test materials and test surfaces.
(3) Be capable of operating from deep cryogenic temperature to a very high temperature.
(4) Exhibit good linearity with minimum creep and hysteresis.
(5) Be single part requiring no mixing or weighing out.
(6) Require minimum curing time.

(7) Require no clamping.
(8) Have a long pot life when mixed.
(9) Be capable of high elongation.

So far no single adhesive has been produced that possesses all the characteristics listed and an adhesive must be selected that has the most appropriate combination of characteristics for the installation to be undertaken.

The following is a summary of the adhesives systems in common use together with their main advantages and disadvantages.

Cyanoacrylate
This material is widely used as a general purpose strain gauge adhesive particularly for stress analysis applications as it is both very fast curing and simple to use. It possesses many of the characteristics of the ideal adhesive previously listed. Polymerisation takes place in less than one minute when the material is spread into a thin film. This can be achieved by the application of thumb pressure, no mechanical clamping system being required. A catalyst is used to control the reactivity rate and is applied to the gauge only but it should be sparingly used. When the relative humidity is low (below 30 %) or the temperature is below 21 °C polymerisation time may be extended to several minutes. This may also happen if a preformed pressure pad is used in place of direct thumb pressure. Normally the bond is ready for use in the time it takes to attach the lead wires (10–15 min).

The adhesive is very simple to use being 'single part' requiring no mixing and when properly handled and stored has a pot life of at least 6 months at 24 °C. To extend the life the adhesive should be stored at about +5 °C but care must be taken to allow the material to reach room temperature equilibrium before the container is opened prior to use to avoid the possibility of condensation forming inside the container which would damage the adhesive and cause a reduction in useful life.

This unique material possesses several other attractive features such as an essentially creep-free performance, an elongation capability in the region of 15 % when fresh (150 000 microstrain) and compatibility with a very wide range of gauge backing materials and test surfaces. Its main limitations are a restricted operating temperature range of −5 °C to +50 °C and the fact that time at the higher temperatures and moisture absorption lead to a deterioration of the bond. Because of these, cyano-acrylates should not be used for long-term installations and should always be adequately protected, particularly when operating in wet environments. Provided this is done cyanoacrylates can be successfully used in wet environments, such as pressure vessels, for short-term tests.

The need to spread the adhesive into a thin film in order to achieve polymerisation may give rise to problems when bonding to very porous surfaces such as poor cast iron and concrete. For the same reasons the material can not be used as a filler on irregular surfaces.

Epoxies and Epoxy–Phenolics

Epoxy adhesives have been extensively used for the installation of strain gauges for many years. A wide selection is available in both room temperature curing form, using an amine catalyst and hot curing form using an acid anhydride catalyst. These are sometimes modified with phenolics for improved performance up to 300 °C and may be available in 'filled' or 'un-filled' form. It should always be remembered that the physical properties of an epoxy or epoxy–phenolic system are largely determined by the type and amount of curing agent used therefore great care must be taken when weighing out the component materials. The temptation to economise by mixing very small quantities should be resisted as this can lead to inaccurate proportions and poor performance.

Cold Cure Systems

Room temperature curing epoxies are primarily used when the component size, temperature restrictions, duration of test programme or the operating environment preclude the use of hot cured epoxies or cyanoacrylate adhesives. Curing is achieved, without the addition of external heat, by the exothermic reaction produced when the hardener and resin are mixed together. Curing time is normally in the order of several hours and moderate clamping pressures of 5–20 psi (35 to 135 kN/m^2) are required to ensure a thin, even glue-line. The addition of external heat will significantly reduce the cure time and normally improve performance. If the adhesive is to be used above the temperature at which it was cured a post cure 20 to 30 °C above the maximum anticipated test temperatures should be carried out whenever possible to achieve optimum adhesive performances and satisfactory first cycle operation. On large structures and components this can be achieved using heat lamps, hot air blowers or various forms of heating tapes. Maximum operating temperatures and elongation capabilities vary depending on the adhesive formulation but 90 °C is a typical limit for operating temperature and 6 % a typical elongation at 24 °C, although some materials are capable of 15 % at 24 °C. Low temperature operation will normally considerably reduce elongation capability.

Hot Cure Systems

These systems will not properly polymerise without the addition of external heat. However, they offer the best all round performance over the widest temperature range from cryogenic to approximately 300 °C for short periods. As with cold curing adhesives, clamping is required during the cure period using slightly higher pressures, typically 15–20 psi (100–200 kN/m^2), and cure temperatures in the region of 150 °C for 2–3 h. As with cold cure systems, best performances will be obtained if the adhesive is post cured 20–30 °C above the maximum operating temperature. The addition of a solvent to the adhesive system reduces its viscosity producing thinner glue-lines that minimise creep, hysteresis, and linearity. The significant improvement in performance over conventional systems has led to the widespread use of solvent-thinned adhesives in high performance transducers, and in stress analysis applications requiring similar performance. Generally, higher clamping pressures are required in the region of 30–40 psi (200–275 kN/m^2) or higher. The presence of a solvent has the effect of very considerably increasing the pot life of the mixed adhesive which can be as long as 6 weeks at 24 °C and extended to 12 weeks if stored at 5 °C. Solvent-thinned adhesives are simple and economical to use, but care must be taken that none of the solvent becomes trapped in the glue-line as this can seriously degrade performance. Manufacturers' air drying periods should be strictly observed to allow complete evaporation of the volatiles. As discussed elsewhere in the chapter particular attention must be paid to the tape handling and clamping system to minimise the possibility of any solvent remaining. Because of the higher cure temperatures and clamping pressures involved, particularly with solvent-thinned systems, the glue-line should be stress relieved by post curing typically 30–45 °C above the cure temperature. This must be done after cooling down from the initial cure temperature and with all the handling tapes and clamps removed.

Although the use of fillers in epoxy systems can increase bond strength, they are not normally used where high performance is required as they tend to produce thicker glue-lines, more hysteresis and creep, and reduce the heat dissipation capability of the installation. However, they are useful in both hot and cold curing forms when dealing with porous or irregular surfaces where a degree of surface filling is required to produce a flat void-free surface prior to installing the gauge. Solvent-thinned adhesives are not suitable for use on most porous surfaces and can not be used as surface fillers.

Polyesters
Polyester adhesives are most commonly used in conjunction with polyester backed gauges but may also be used with epoxy and polyimide types. Although polyesters possess the high shear strength required of a good strain gauge adhesive, many exhibit a very low peel strength that may give rise to handling problems during the installation process particularly when removing handling tape following bonding when delamination may occur. Hence some polyesters currently available should not be used for high elongation or impact work. However, one advantage of some polyesters is their ability to operate at temperatures up to $+150\,°C$ after cold curing at only $+5\,°C$ without a subsequent post cure stage. This can be a major advantage when dealing with some stress analysis applications requiring operation above room temperature but where hot curing would be difficult or not permitted.

Phenolics (Bakelite)
This group of adhesives was once widely used particularly when bonding Bakelite-baked gauges. However, the material forms volatiles and water vapour during the cure period and therefore high clamping pressures are required and an adequate escape path must be provided to ensure proper egress of both these by-products from the glue-line to prevent the formation of water droplets and gas bubbles. Because of the high clamping pressures and the long cure times involved phenolics have largely been superseded by epoxy and epoxy–phenolic adhesives which are easier to use and can produce a superior performance.

Polyimide
Polyimide had been in use for some years as a gauge backing material and more recently has been made available as an adhesive. Because of the formation of volatiles during the cure, with the resulting risk of bubbles occurring in the glue-line, polyimides have proved more difficult to use than epoxy–phenolics. However, the material has good potential as a high temperature adhesive and may become more widely used in the future.

Nitro-Cellulose
This type of adhesive sets by solvent evaporation and once was very widely used, however it requires the gauge backing to be porous to allow the evaporation to take place. Most modern strain gauges have non-porous backings therefore its use is restricted to paper backed gauges. The material is hygroscopic and must be protected once the adhesive is *fully* cured. As

about 80 % of the adhesive evaporates during setting a residual tensile stress results which may require releasing. Caution should be exercised when removing flux as many flux removing solvents will damage the adhesive layer also the solvent in the adhesive itself may damage some materials. This adhesive allows easy removal of an installed gauge without damage to the surface and can operate up to 84 °C but drying times are usually in excess of 24 h.[1]

Ceramic Cements

These cements are used for the installation of strippable-backed or free-element gauges where the avoidance of all organic materials such as the adhesive and gauge backing is necessary to permit use of the installation up to 700 °C for continuous operation and 800 °C for short-term excursions. The cements are normally based on aluminium phosphate and silicon in a water-based slurry. Modern developments of ceramic cements are available as single part materials which can be applied using a small brush, air dried and baked, resulting in a hard but porous coat. Although the technique is basically simple and requires no special equipment the installation method is not suitable for beginners as the degree of skill required is higher than for a conventional room temperature application, and practice is necessary to avoid significant scrappage.

It should be remembered that maximum elongation of the adhesive may be only 0·5 % or less. The material can also be used as a protective coating for a conventional organic installation in order to extend life at elevated temperature. One example of this is to provide protection from hot gas flow such as might be encountered when gauging an exhaust valve stem for an IC engine where the metal temperature is acceptable for a conventional installation but would be severely limited by the hotter gas flow.

Flame Sprayed Alumina

This system is an alternative method for installing free-element gauges for high temperature work, such as on gas turbine components, and involves the spraying of droplets of alumina or other oxides on to the test surface to provide an insulating layer followed by application of the gauge which is achieved by spraying additional layers of alumina. Wire gauges are frequently used with this technique as the round wires are more easily embedded and they are better able to withstand the spraying forces involved. They are often supplied attached to a removable carrier which is used to assist handling and to keep the gauge grid flat whilst the initial layer of alumina is applied. The carrier can then be removed and the spraying completed. The carrier technique can also be employed when installing gauges with ceramic cements. The major problem with the flame spray method is that it requires special equipment and an experienced operator. Special guns similar to those used for metal spraying are used with alumina, in either rod or powder form, being fed into the gun where it is melted into droplets by an oxy-acetylene flame and propelled to the test surface by the gas plasma. In addition to its high temperature performance (up to 1000 °C for some dynamic measurements) the technique also offers a method that requires no cure cycle, a reasonably quick installation time together with better mechanical and electrical properties than ceramic cements, and a coating that is suitable for use in both vacuum and radiation environments.

SURFACE PREPARATION

A universal surface preparation procedure that has been tried and proven over many years and which can equally well be used for on-site laboratory applications consists of five basic operations (these may be slightly modified for compatibility with different test materials).

(1) Solvent degreasing.
(2) Abrading.
(3) Application of gauge layout lines.
(4) Conditioning.
(5) Neutralising.

The method may be regarded as a standard procedure regardless of the type of adhesive or strain gauge to be used.

Solvent Degreasing

The object of this first operation is to remove as far as possible organic contamination, oils and greases etc. from the surface. An area considerably larger than that required by the strain gauge should be prepared to allow for the subsequent application of a suitable protective coating and to minimise the risk of recontaminating the gauge area as cleaning proceeds. Degreasing can be carried out using several methods and solvents but it should be noted that some materials react to chlorinated solvents. For site and lab work there is much to recommend the use of aerosols as these are convenient to use and easy to transport in a strain gauge application kit. Their use also ensures that the bulk material can not be contaminated by used material. Aerosols of Chlorothene NU and Freon TF are suitable and widely used for this purpose. Ultrasonically agitated baths using materials such as 'Arklon', isopropyl alcohol, 'Geneklene', chlorothene etc. are useful for small components particularly in a production environment such as the transducer industry and, like hot vapour baths which may also be used, offer the advantage of complete component degreasing. Some caution is, however, needed when using solvents to degrease plastics. As is well known some plastics can be severely damaged by solvents, and it should be carefully verified with the manufacturer of the plastic, which solvents are safe to use. However, many of the plastics commonly used in engineering today may be safely degreased using either Freon TF or isopropyl alcohol. Porous surfaces may also present difficulties, and materials such as cast iron and titanium may require heat to drive off absorbed hydrocarbons or other liquids such as cutting fluids, lubricants etc. A suitable procedure for cast iron or titanium is to heat for 2–6 h at 175 °C before degreasing.

Good surface preparation is essential in order to achieve the optimum conditions for good, strong bonds. It is a step that should never be treated in a less than thorough manner if consistent, high quality bonds are to be achieved.

The primary objectives are to bring the test surface to a chemically clean state with an alkalinity corresponding to an approximate pH value of 7 so that the adhesive can properly wet the surface and bond to it. In addition the correct degree of roughness must be obtained appropriate to the application requirement and the location provided with visible layout lines for accurate alignment of the gauge.

It cannot be overstressed that throughout the process the use of clean, fresh and uncontaminated materials is vital. Hands should be cleaned and kept clean during the procedure. Similarly, it is important that the general environment in which the gauging is to take place should be as clean as possible and free from air borne contamination.

Clean room conditions are not required for general stress analysis work although they are used by some companies in the high performance transducer industry. A less expensive way to reduce the airborne contamination problem in a workshop or lab area is to make use of 'Laminar flow' cabinets which provide a curtain of filtered air across a working area, however, this approach has its limitations when dealing with large components or structures.

On-site conditions can be particularly difficult to deal with especially when open to the weather, however, even in these circumstances it is often possible to improve the local environment by erecting protective screens or, if the structure allows, by constructing a cabinet around the gauge area to keep the worst contamination away. Even a crude solution such as a large cardboard carton turned on its side to form a cabinet can be effective when extremely poor conditions prevail. If poor conditions are known about in advance thought should be given to the use of gauges with special options and adhesives and protective coatings that require no mixing or curing. The weldable gauge and butyl rubber protective coating systems are good examples of materials intended for use when the installation environment is poor.

Surface Abrading
Following degreasing, the component surface must be brought to the correct degree of surface finish at the gauge location. Many modern strain gauge adhesives are formulated to produce glue-lines typically of the order of 0·0001 to 0·0002 ins thick (0·0025 to 0·0050 mm); therefore strain gauge adhesives generally require a relatively smooth, matt, surface finish which also has the advantage of being easy to clean. The exact finish required will depend on the type and purpose of the installation and the adhesive to be used. As a guide, a stress analysis installation using a cyanoacrylate or medium viscosity epoxy (i.e. no solvent thinning) will require a surface finish of approximately 63–125 microinches r.m.s. (1·6–3·2 micrometres r.m.s.).[2] Efforts to improve adhesive performance, particularly in the area of creep, have led to the development of solvent-thinned materials which are capable of even thinner glue-lines, consequently a slightly better surface finish is desirable to obtain the best performances from such materials. In these cases the finish should be between 16–63 microinches r.m.s. (0·4–1·6 micrometres r.m.s.). The need to take measurements involving high elongation and plastic strains requires a coarser surface and better results will be obtained if the finish is around 250 microinches r.m.s. (6·4 micrometres r.m.s.) taking the form of cross hatching on the component surface. Use of a ceramic type adhesive with 'free-element' gauges also calls for a surface finish of around 250 microinches r.m.s. in cross hatch form.

In some cases the surface may require to be 'improved', using conventional engineering methods such as grinding, disc sanding, hand filing etc. to remove mill scale, machining marks and other surface imperfections before final surface preparation can begin. Where possible surfaces plated with cadmium, chrome etc. should have the plating removed in the gauge area. Many types of plating are subject to creep therefore if removal is not permitted due consideration of the possible effects on gauge performance should be given. All paint should be removed from the gauge location. The final surface texture can be achieved by wet abrading with an appropriate grade of silicon carbide paper. As a general guide 280 grit is suitable for steel components and 400 grit for softer material such as aluminium and copper.

Silicon carbide paper is preferred as the grit is securely bonded to the backing paper reducing any tendency for the grit to drop out during use. As abrading proceeds the paper becomes loaded with debris from the surface producing a progressively finer finish. Whenever possible it is preferable to wet abrade using a weak phosphoric acid 'conditioner' in conjunction with the silicon carbide paper. Conditioners generally accelerate the cleaning process and act as a gentle etchant on some materials and should *never* be allowed to dry on the surface, additional liquid being added as required during abrading in order to keep the area wet. Failure to do this may cause contaminants to be driven into the surface.

A circular abrading motion, producing multidirectional marks, should be used for all installations with the exception of high elongation, where a cross hatch finish produces better results.

When abrading is complete the surface should still be wet. The liquid and residue can then be wiped from the surface using either a gauze sponge or tissue pad. Care should be taken to start the wiping operation just inside the abraded area, to avoid dragging in any contamination from the unclean boundary then, using firm pressure, a single slow wipe should be taken through the area, followed by a single slow wipe in the opposite direction using a clean pad and again starting just inside the abraded area.

Wiping with a vigorous back and forth motion should be avoided as it may lead to recontamination of the abraded area.

There are some materials on which phosphoric acid based conditioners should not be used most notably magnesium, natural or synthetic rubbers, and wood. In these cases abrading must be carried out dry or in conjunction with neutraliser. Magnesium should not be abraded to produce fine particles, instead the location should be scraped with a knife, or filed.

Overhead or vertical surfaces also present problems when wet abrading and in some cases it may be preferable to dry abrade followed by scrubbing the surface with conditioner using cotton applicators or gauze sponges. This may need to be repeated several times until there is no indication of contamination. Final drying should be as described for wet abrading.

Marking the Gauge Location

Accurate alignment of the strain gauge on the component is nearly always a requirement. The normal method of achieving this is to provide two crossed lines on the surface at the point where the measurement is to be made. When the gauge is installed its alignment marks can be orientated to coincide with the reference lines on the component surface. However, it is important to avoid scoring the surface giving rise to burrs that may both damage the gauge and cause stress concentration. Marking should take the form of a burnished line and may be achieved on soft material by using a 4H pencil. Harder materials may require the use of a ballpoint pen or similar tool. Graphite pencils should never be used on high temperature alloys where the operating temperature may give rise to carbon embrittlement. For production work such as may be met in the transducer industry it may be possible to avoid the physical marking of the component by projecting reference lines optically on to the surface, or by arranging for the gauge backing to align automatically with some datum on the component. The latter method will normally require special trimming of the gauge backing.

Surface Conditioning
Following the marking of reference lines on the component the surface must be repeatedly scrubbed with conditioner, using cotton applicators, to remove any contamination caused by the operation. This process is continued until a clean applicator tip is no longer discoloured by the scrubbing. As before the surface of the component must be kept flooded with conditioner and not allowed to dry out. Final drying is achieved by wiping a gauze sponge or tissue through the gauge area as previously described. Again care should be taken to wipe inside the prepared area to minimise the risk of recontaminating it from the surrounding surface.

Neutralising
The preceding steps involving the use of a conditioner will have left the surface slightly acidic and some adhesives, particularly cyanoacrylates, will not bond to an acidic surface, therefore the final step is to bring the surface alkalinity to the optimum condition of a pH value of around 7. This can be achieved by scrubbing the surface using a cotton applicator in conjunction with a neutraliser, usually a water based ammonia solution.

As with conditioning, scrubbing should continue until a clean applicator tip is no longer discoloured. Care should be taken to restrict the scrubbing to within the abraded area and, as with conditioner, the neutraliser should not be allowed to dry on the surface, fresh material being added as required. Final drying should again be carried out as previously described using a gauze sponge or tissue. Once neutralised the surface is ready for positioning of the gauge and the application of the adhesive which should be carried out as soon as possible before the surface re-oxidises. The allowable delay between surface preparation and bonding varies depending on the material but in general steel should not be left for more than 45 min, aluminium, beryllium or copper 30 min and titanium for not more than 10 min. At all times care should be taken not to touch the cleaned surface with fingers.[2]

GAUGE PREPARATION

Most modern foil strain gauges require little, if any, pretreatment prior to the bonding operation, however there are still gauges available that require their backings to be abraded or cleaned and in these cases the manufacturers' instructions should be followed. Gauges should never be contaminated by being handled with the fingers. Touching the foil element can lead to long-term changes in resistance, and touching the backing can lead to bonding problems. The gauge should always be handled by holding the gauge backing as far from the foil as possible using a pair of clean, round nosed, tweezers. It is important that the foil is not scratched or marked as this can seriously degrade the gauge performance. It is as well to remember that modern gauges use foil only 0·0001 to 0·0002 in thick.

Mounting

In order to assist in handling and positioning, the unbonded gauge is normally mounted on a suitable adhesive tape. The type of tape and how it is used will depend on the combination of strain gauge, adhesive and cure temperature to be used. Generally a cellophane tape can be used for cold curing applications but a mylar tape must be used for any installation to be hot cured because cellophane tape hardens on exposure to temperature becoming difficult to remove and increasing the risk of damage to the foil element. The mounting operation should be carried out on a rigid surface that can be easily cleaned. A small square of glass makes an excellent surface that is readily cleaned by scrubbing with cotton applicators and neutraliser.

There are two methods of applying the handling tape. Which one is used depends on the construction of the gauge and the adhesive selected. The first should be used if the gauge is of the type possessing low peel strength between backing and foil. In this case the handling tape should be applied to the edge of the backing only and *not* across the foil area. If this precaution is not observed delamination of the foil may occur when the tape is removed. This method also has advantages when solvent-thinned adhesives are being used as the absence of tape around the edges of the gauge backing ensures

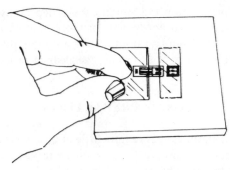

Fig. 1.   Method of applying handling tape to the edge of the gauge backing only.
This method is preferable when installing gauges with low peel strength between foil
and backing, or when using a solvent-thinned adhesive. Note the use of a clean piece
of acetate sheet to help keep the gauge in position on the mounting plate. On no
account should the fingers be used to touch an open faced gauge. Courtesy of Micro
Measurements.

that no excess solvent, air or surplus adhesive can become trapped under or
around the edges of the gauge (Fig. 1). Trapped solvent can be a serious
problem and may cause weakening on the glue-line, the formation of voids
in the glue-line and delamination of the foil from the backing. The presence
of voids under a gauge grid will also cause hot spots to develop when a
current is passed through the gauge and lead to instability. This method of
tape handling allows all the exposed faces of the gauge to be coated with
adhesive, including the grid, so that after curing the gauge is completely
encapsulated offering a high degree of protection and this is frequently used
in the transducer industry. Not all adhesives can be used in this way and the
gauge must either be fitted with solder dots which protrude through the
adhesive layer or the handling tape must cover the part of the gauge tab that
will eventually be tinned with solder (Fig. 2).[3]

The second method is to apply the tape across the whole foil area. This
makes it easier to handle and position the gauge but can only be used when
the gauge foil-to-backing peel strength will allow. If solvent-thinned
adhesives are to be used the tape should not be wider than the gauge backing
to avoid entrapment problems, this also makes removal from the
mounting surface easier. If a cyanoacrylate adhesive is to be used it is
permissible to use a wide tape in order to protect the operator's fingers from
adhesive spread. A wide tape may also be used with other types of adhesive
provided they do not contain solvents. If the gauge to be used has been
supplied cleaned and treated for bonding it can be removed from its

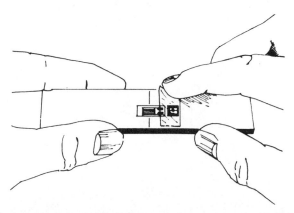

FIG. 2. Mounted gauge positioned on specimen prior to application of the adhesive. Courtesy of Micro Measurements.

packaging and placed on the cleaned mounting surface with the foil side facing up. The tape can then be attached to one side of the mounting surface and slowly rolled towards the gauge until the edge of the backing is attached. Care at this stage will avoid problems caused by the gauge moving about under the influence of static electricity. Once the backing edge is attached the process may be completed by gently rolling the tape across the gauge grid or the backing edge as appropriate. Care should be taken not to use excessive pressure that might cause the tape and gauge to be stretched as this can result in a high installed resistance.

To remove the gauge and tape assembly from the mounting surface the tape should be lifted at a shallow angle until the gauge is free. Under no circumstances should the assembly be removed by doubling the tape back on itself (Fig. 3). Gauges with brittle backing will not survive such treatment and all gauges may suffer damage to the foil leading to unacceptable installed resistance values.[4]

Use of Terminal Strips

For most stress analysis installations the use of printed circuit terminals is advisable. These are available in a number of configurations constructed from electrodeposited copper on a flexible Teflon backing which has been treated for bondability. Fibreglass backings are also used particularly for cryogenic applications, however, they are not as conformable as the Teflon types.

The primary function of a terminal strip is to provide an anchor and a

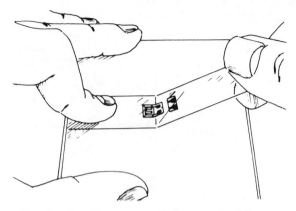

FIG. 3.   Handling tape applied across the foil area.

junction between the main lead wire and the relatively small and delicate
jumper wires to the strain gauge. This method prevents damaging forces
being transmitted to the gauge via the lead wire as might occur if the lead
wires are accidentally subjected to a large force such as someone tripping
over them! The lead wire should fail at the terminal strip leaving the gauge
undamaged. Use of a fine jumper lead requires a smaller solder connection
on the gauge tab which both reduces the risk of delaminating the tab during

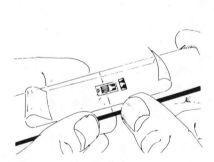

FIG. 4.   Mounted gauge positioned on
specimen.

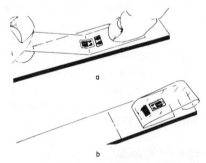

FIG. 5(a).   Tape assembly being lifted
at a shallow angle, away from the
specimen surface. (b) When gauge and
terminal are free of the surface the loose
end of the tape is tucked under and
pressed to the surface in preparation
for the application of the adhesive.
Courtesy of Micro Measurements.

the soldering operation and increases the gauge's fatigue life. As copper has an inherently poor fatigue performance care should be taken when using copper terminals in high strain regions. To avoid fatigue failure in the centre section of a standard 'dog bone' type terminal, which would break the electrical circuit, the terminal should be cut in half, taking care not to burr over the copper, and just one half used as an anchor for both leads (Fig. 10).

Terminal strips can be bonded with the same adhesive as the gauge and can therefore be mounted on the handling tape at the same time. As terminal backings are usually thicker than the gauge backings a gap of approximately 0·062″ (1·5 mm) should be left between the two to reduce the build-up of adhesive that would occur if the backings were butted together and which would lead to bond failure. This is particularly important when using cyanoacrylate adhesive or when undertaking high elongation measurements. (Figs 4, 5 and 9).[5]

CLAMPING

As already noted the majority of strain gauge adhesives require pressure in order to obtain their correct glue-line thickness and assist set-up of the adhesive. In the case of those that set or cure in one or two minutes thumb pressure may be sufficient although the human thumb limits the size of gauge that can be covered in this way. Most adhesives require much longer cure periods therefore a mechanical clamping system is required. The basic components of a system that can be used for most adhesives and gauges consist of a thin Teflon sheet applied over the gauge followed by a silicone rubber pad and a stiff backing plate to which the clamping force is applied (Fig. 6). The Teflon film, typically 0·003 in thick, has two main functions. First, because of the materials low coefficient of friction it can be used as a slip sheet between the gauge and the clamping assembly to allow some movement of the rubber pad and back-up plate when the clamping force is applied without disturbing the alignment of the gauge. Its second function is to provide a non-stick interface between the rubber pad and gauge particularly when the handling tape has been applied across one edge of the backing leaving the grid exposed and which may have been coated with adhesive. As previously mentioned this technique offers advantages when using solvent-thinned adhesives. The film should be cut larger than the gauge backing and rubber pad and may be anchored in place with a suitable tape.

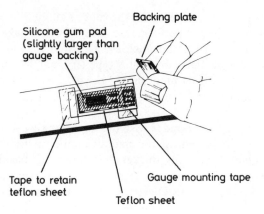

FIG. 6(a). Standard clamping system for gauges encapsulated with adhesive at the time of installation. This method is also recommended when solvent-thinned adhesives are used. Courtesy of Micro Measurements.

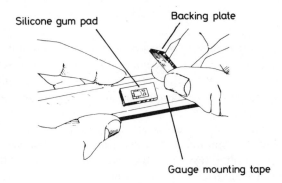

FIG. 6(b). Standard clamping system used when handling tape is applied across foil area. Courtesy of Micro Measurements.

The silicone rubber pad acts as a pressure equaliser between back-up plate and gauge ensuring an even glue-line. Although other types of synthetic rubber can be used as pads, silicone rubber of about 3/32 (2·5 mm) thickness has been found to be a very suitable material primarily because it retains its resilience at high cure temperatures and can be re-used many times to offset its high initial cost. The pad should be cut only slightly larger than the gauge backing (and terminal strip if used) to assist proper spreading of the adhesive, reduce the possibility of trapping solvents and to ensure that the clamping force is concentrated on the gauge. Flat silicone rubber sheet is very flexible and suitable for the majority of installations. However, when difficult surface contours are encountered a cold curing silicone rubber can be used to make a cast of the surface. Plasticine makes a suitable dam on the component to retain the liquid rubber whilst it cures. Making a cast may allow a flat or less complex backing plate to be used. The backing plate should simply be stiff enough not to deform under the clamping force and should match accurately the contours of the components surface, taking into account the thickness of the rubber pad. Wherever possible the clamping force should be applied by a spring or dead weight system. This principle is particularly important when hot curing is involved to ensure that a constant force is applied throughout the cure cycle (Fig. 7). If the force is applied using a screw thread system such as a 'G' clamp or pipe clip there is a risk of thread relaxation or differential expansions during the temperature cycle resulting in a reduction of the force being applied. Screw thread systems may, however, be of use in cold curing application where the compression of the silicone rubber may be used as an indication of the force being applied. Pressures required vary depending on the type of adhesive, and manufacturers' recommendations should always be followed. Values from 5–50 psi (35–300 kN/m^2) are typical for epoxy and epoxy–phenolics. Many stress analysis applications involve 'one-off' installations, often on large or fixed structures where special purpose fixtures may be uneconomic or difficult to manufacture. An installation may also need to be completed quickly allowing insufficient time for fixtures to be made. In these cases a great deal of ingenuity must be exercised to provide a solution to a particular clamping problem. Too many

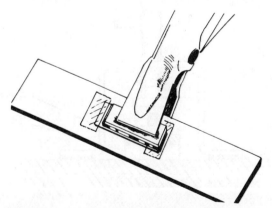

FIG. 7.   Clamping using a spring clamp.

possibilities exist to make a comprehensive list in this chapter but the following examples may help in showing what can be done.

The pipe clip can be used with considerable success on many types of circular components including shafts and pipe work, and can be considerably improved for hot curing application if modified to incorporate a spring system. Large flat areas or large internal and external diameters can present problems. However, a spring operated clamping assembly mounted in a suitable piece of top-hat section material can solve all three problems. The top-hat section can be attached to the test surface astride the gauge area using a rapid curing adhesive, which will allow easy removal when the installation is complete (Fig. 8). Another approach is the use of a vacuum bag. This can be easily constructed by placing strips of vacuum putty around the installation area which is then covered with a layer of mylar film or similar material having first inserted a suitable length of tubing for connection to a vacuum pump. A pressure of one atmosphere can be applied to the installation by this method which is generally enough for cold curing epoxy systems.

A cruder method that could be used in an emergency involves using a piece of plastic tubing, similar to that used for a garden hose. After placing on the backing plate the tubing can be collapsed by stretching a length of tape over the tube. Sufficient force can be generated for some adhesive systems. Magnets may also be successful on ferrous components.

Installing gauges in holes of small to medium size can present problems. However, these can often be solved by use of cold curing silicone rubbers. For a through hole a plug can be cast slightly longer than the hole. A bolt is

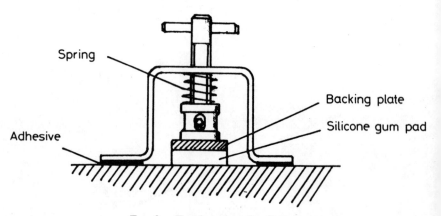

FIG. 8.   Top hat clamping fixture.

passed through the centre of the plug and two suitable washers used as end plates so that on tightening a nut the silicone rubber is squeezed thus expanding it to become a tight fit in the hole. Blind holes may be treated in a similar way using a washer and dead weight.[6] A possible solution for long through holes or deep blind holes is to use or modify such items as football bladders and bicycle inner tubes. Repetitive applications of gauges to identical pieces of hardware may make it worth while constructing special jigs and fixtures to ensure consistent application of the correct force and to facilitate rapid assembly and disassembly of the clamping system. In addition to the incorporation of dead weights or springs on such fixtures the more sophisticated production tools may employ pneumatic or hydraulic principles.

CURING

Curing requirements vary depending on the type of adhesive used and the manufacturer's instructions should be carefully followed to achieve optimum performance. However, a few general points are worth observing.

When hot curing it is important to establish that the manufacturer's recommended cure temperature has been reached and that the correct temperature/time schedule is achieved. If the component is large it may take some considerable time to reach oven temperature making it necessary to monitor the temperature at the gauge location using thermocouples or temperature sensors rather than relying on the oven temperature indicating system. Hot curing on large or fixed structures normally requires the use of alternative heat sources such as heating tapes or heat lamps to achieve the cure temperature making it essential to monitor component temperature to avoid undercuring and the poor adhesive performance that would result. Problems such as bubbles, uneven glue-lines and high residual stress in the glue-line can result from placing components into an already hot oven with the result that the manufacturer's recommended heating rate is exceeded. Wherever possible curing should start in a cold oven. A typical heating rate for a solvent-thinned epoxy–phenolic is in the region of 3–11 °C per minute. If circumstances dictate the use of an already hot oven the component should be placed in some sort of insulated box, known as a 'lag box', to slow down the heating rate. After completion of the cure the component should be allowed to cool naturally before the clamps are removed. As previously mentioned after a cure involving heat and pressure the glue-line may have locked in residual stresses. To achieve maximum long-term stability these stresses must be relieved by the process of post curing. This can be done by removing all the clamping material and tape after the initial cure and placing the component back into the oven. To ensure that the adhesive reaches the right viscoelastic state to release the stresses the post cure temperature should exceed the cure temperature by 20–30 °C. Post curing will also improve an adhesive that has been cured below the expected maximum operating temperature. Then the glue-line temperature during the post cure should exceed the expected maximum operating temperature by 20–30 °C. The performance of many cold curing adhesives can be improved by a subsequent hot post cure. Curing for longer than the schedule given by the manufacturers will not normally give rise to any problems. Much more serious is the possibility of under curing.

LEAD WIRES

Before considering the attachment of lead wires to the bonded strain gauge some comments on techniques and selection are appropriate.

The lead wire can be greatly influenced by the operating environment and give rise to significant errors in the gauge output. This is particularly true whenever the lead wire forms part of the measuring circuit, as it does in a quarter bridge or half bridge system. The effects of the environment on the lead wire must therefore be properly understood to enable the correct selection to be made (see also Chapters 4 and 5).

Essentially three decisions have to be made.

(1) The type of insulation to use.
(2) The type of conductor to use.
(3) The form of cable construction to use.

INSULATION

Cables insulated with many types of material are available and the following lists some of the more widely used materials together with their limitations and advantages.

For a large number of installations the operating temperature will be the overriding factor affecting choice, however, other factors may also be important such as ability to survive underwater or to resist the effects of chemical attack etc.

Polyvinyl Chloride (PVC)

PVC is generally inexpensive and it is the most widely used material for strain gauge lead wires. It gives satisfactory performance over the temperature range $-50\,°C$ to $+70\,°C$, however, the cheaper grades can suffer from pin holes therefore lower quality PVC insulated wire should not be used for external applications or underwater. However, high quality PVC can be used in such circumstances although for critical installations a quality control check should be made by immersing coils of the selected material in a water filled tank for several hours. The ends of the conductor should hang down outside the tank below the water level and periodic checks of insulation resistance between conductor and water made to identify any unsuitable material. The better quality material may even be used in water under pressure.

PVC should be degreased with a suitable solvent and primed with a nitrile rubber solution over the length in contact with the protective coating in order to promote better adhesion between coating and lead wires. The material can be obtained in a wide variety of colours and forms of construction but it is easily damaged by heat and solvents and becomes brittle at low temperatures.

Polytetrafluoroethylene (PTFE)

PTFE (Teflon) is the most commonly used material to cover the temperature range outside the limits of PVC, being able to operate between $-260\,°C$ and $+260\,°C$ and underwater, however, the material requires surface etching to produce a good bond with the protective coating. Like PVC the material is available in a wide range of colours and forms of construction but can be difficult to strip without damage to the conductor. Hot wire stripping is recommended. The material can cold flow at room temperature under pressure. For underwater use, an immersion test should be made first as for PVC.

Polyimide

This material can provide an extremely tough insulation and therefore is often useful when abrasion is likely. It will operate between $-269\,°C$ and $+315\,°C$, however, the material cannot be coloured and an additional wrapper, usually PTFE, has to be used to provide a colour code. The material possesses excellent resistance to radiation and to outgasing in high vacuum. Although basically flexible some forms of construction can be springy and therefore difficult to handle in confined spaces. Because of its inherent toughness stripping without damage to the conductors can be difficult. Light abrading to remove surface gloss is recommended before bonding to the coating system. Polyimide solid conductor wires cannot be stripped using a soldering iron. Multiple conductor cables are available in several types of construction.

Fibreglass
Fibreglass braid insulation, capable of operating between $-269\,°C$ and $+480\,°C$, is not normally colour coded and is liable to fray around cut edges. Probably the most difficult problem with this material is its tendency to absorb flux-contaminated solvent during the soldering operation, the problem being aggravated by the fact that fibreglass insulation is frequently used in association with the more corrosive fluxes used with high melting point solders.

Mineral Insulation
Mineral insulation is mostly used for high temperature and long-term installations on fairly substantial structures and uses either a copper or stainless steel sheath packed with magnesium oxide as the insulator. Whilst being expensive and requiring special tools it does give a completely sealed, water-tight cable run very resistant to mechanical damage.

CONDUCTORS

Copper is still the obvious first choice, usually tinned, and in both solid and stranded form. Whilst light gauge wires are required for direct connection to the strain gauge, heavier gauge wires should be selected for long lead wire runs to reduce lead wire resistance and the resulting desensitisation of the installation. For the higher temperature ranges the copper should be silver or nickel plated. Single strand conductors yield a lower resistance per unit diameter and lend themselves to spot welding, however, stranded conductors perform better under vibratory conditions. For use with high temperature strippable-backed gauges the conductor may take the form of a nickel–chrome ribbon to allow easy spot welding to the gauge tab. Insulation is provided by a fibreglass braided sleeve.

Cable Construction
A strain gauge installation's immunity from electrical noise is largely determined by the type of cable construction used and the choice of cable is therefore a very important decision.

Ribbon cables, using PVC, PTFE and polyimide insulation, are generally both inexpensive and convenient to use, being easy to attach to the test surface. However, whilst this form of construction is perfectly satisfactory in low noise environments it is not suitable when electrical noise is present in either electrostatic or electromagnetic form.

If electrostatic noise is the main problem a shielded cable is the most effective answer using a braided wire or conductive foil shield. As ribbon cables are not normally supplied with a shield a conventional round twisted cable will probably have to be used.

Crosstalk between conductors in a multiple conductor cable may become a problem if cables are long, e.g. over 15 m. This can be reduced using a form of construction composed of individually shielded pairs, one pair being used for excitation and one for the signal.[7]

As the strength of a magnetic field varies depending on the distance from the source of the field, magnetically induced noise can be minimised by twisting the conductors of the cable together. This method ensures that the noise voltage induced into each conductor is essentially the same because the average distance of each conductor from the source is approximately the same. When the signal is amplified the noise induced voltage is cancelled provided the amplifier used possesses good common-mode rejection characteristics.

Specially woven cables may be needed to counter very severe electromagnetic noise. These are designed to minimise common-mode voltage by elimination of the spiral inductive loop. This is achieved by the use of four conductors in place of the normal twisted pairs, each of two wires being connected in parallel.[8]

Lead Wire Attachment Techniques for Stress Analysis

For stress analysis applications the three lead system, as described in a following chapter, is the most widely used method of connecting self-temperature compensated gauges to the instrument (Fig. 9). As already mentioned in this chapter the use of terminal strips offers several important advantages and they should be used in all cases unless the strain gauge has been specially constructed to permit the direct attachment of the main lead wires.

The jumper wire previously referred to under the use of terminal strips can either be a separate conductor of fine gauge copper or alternatively a single strand of the main lead wire can be continued through the terminal strip to the gauge. A strain relief loop should be formed in the jumper wire to avoid a tight connection between the gauge tab and terminal strip. The

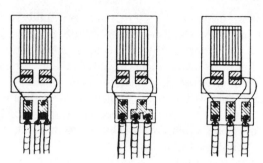

FIG. 9.   Various methods of making three-wire cable connections. Courtesy of
Micro Measurements.

loop can be formed in the same plane as the gauge as shown in Fig. 10 but
care must be taken to keep the loop on the gauge backing to reduce the
possibility of grounding. This is a particularly important point when using a
strand of the main conductor which will not normally be provided with
insulation. When gauges with integral jumper leads are being used it may be
more convenient to form the strain relief loop in the vertical plane (Fig. 11).
Particular care must be taken for high strain applications and when the
maximum fatigue life of the strain gauge is required.[5] If the completed
installation is to be operated under the influence of a high magnetic field
strain relief loops must be avoided to minimise the effects of magnetically
induced noise.[7]

When dealing with installations other than those intended for stress
analysis it may be necessary to connect a number of gauges together to form

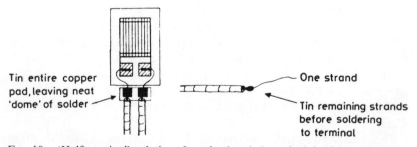

FIG. 10.   'Half-terminal' technique for using bonded terminals in high strain fields.
(Note: when the main lead wire is stranded, it is often convenient to cut all strands
but one to fit the size of the copper pad. The long strand can then be used as the
jumper wire. Soldering is made considerably easier by this method.) Courtesy of
Micro Measurements.

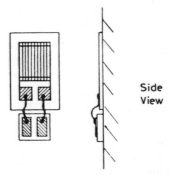

FIG. 11. 'Vertical' stress/relief loop. Often used when gauges have integral jumper leads. Courtesy of Micro Measurements.

a complete Wheatstone bridge as would normally be the case when producing a transducer. In these cases a fine copper wire, similar to that used to form jumper wires, can be used to connect the gauges together and to provide a link between the gauges and a suitably placed terminal strip. Solid copper wires in several diameters are readily available using polyurethane or nylon insulation which is easily removed by applying heat from a soldering iron. Polyimide insulation is also available and although difficult to remove provides excellent abrasion resistance and is particularly useful when the lead wire path involves sharp corners.

LEAD WIRE ATTACHMENTS

Three methods are commonly used to attach lead wires to strain gauges, soft soldering, silver soldering and spot welding. Which method is used is decided by the operating temperature of the installation.

Soft Soldering

Soft soldering is the most widely used method with the ability to cover the temperature range from deep cryogenic to $+300\,°C$ by selecting the appropriate alloy. By far the most widely used alloy is a 63 % tin, 36·65 % lead, 0·35 % antimony combination with an operating temperature range from cryogenic to $+183\,°C$. The addition of a small amount of antimony overcomes the problem of phase transformation of the tin content (tin disease) when operating tin/lead solders for long periods at cryogenic temperatures. Some of the higher melting point alloys are more difficult to handle having slightly inferior wetting and flow characteristics therefore a solder with an operating temperature range to match the installation requirements should be used. Where possible eutectic alloy's should be used as these have virtually no 'pasty' period between the solid and liquid states, reducing the possibility of erroneous data from a partly failed joint.

Consideration may need to be given to factors other than temperature when selecting a solder, such as fatigue life, electrical conductivity, and possibly the type of lead wire to be soldered. Gold leads as found on semi-conductor gauges should not be soldered with high lead content alloys as the lead 'dissolves' the gold leading to early failure of the joint.

Soldering is often approached with some apprehension by the beginner, particularly when faced with miniature gauges, and as a result it frequently causes problems. Whilst some skill is required to produce sound joints this is easily obtained with a little practice. In recent years manufacturers have developed gauges with special options, intended to simplify lead wire attachment, including integral terminal strips, pretinned gauges and pre-attached fly-leads. However, many installations will be met when these options are of little help and may even make the installation more difficult. Pre-attached leads are often requested by users particularly on miniature patterns, but it should be remembered that many are not insulated, short in length and may give rise to uneven glue-lines because the clamping pressure during curing cannot be uniform. They can also be difficult to handle in confined spaces and in some forms are easily damaged. Also the user has no choice over the lead wire position on the gauge tab which again can be important in confined spaces and in high strain locations where the lead wire orientation on the tag may affect the fatigue life obtained.

Flux

All fluxes are corrosive and any residue remaining after the soldering operation will lead to subsequent degradation of the installation in the form of permanent zero shift due to its corrosive action on the grid (increase of gauge resistance) or by providing a conductive path-to-ground which effectively shunts the gauge (decrease in gauge resistance). For these reasons the minimum of flux should be used followed by very thorough removal. A rosin cored solder will be found satisfactory for constantan foil and no additional flux need be used. Other types of foil such as nickel–chrome and isoelastic can be more difficult to solder and the use of an additional flux may be necessary. To assist soldering on these foils options such as pretinned tabs or pre-attached leads are available. When using the appropriate solvent to remove flux particular attention should be paid to the area around the lead wire insulation and to the possibility of flux having 'wicked' along the lead wire. Special treatment may be needed if wicking has occurred to ensure complete removal of the flux to avoid contamination of the protective coating and gauge grid.

Soldering Equipment

There is a tendency to use miniature soldering bits, however, these may not always be helpful as the properly bonded foil strain gauge is an efficient dissipator of heat and can rapidly cool small tips leading to difficulty in forming a satisfactory joint.

A good general purpose iron for strain gauge use should be in the 25 watt range with a nickel-clad chisel or screwdriver shaped tip 1/16″ wide. It is also an advantage if the iron tip temperature can be adjusted to suit the many different solders and conditions that may be met from indoor laboratory work to cold weather site work. The control of tip temperature is also an advantage when working on materials such as plastics which are easily damaged by an excess of heat.

A Technique for Soldering to Open Faced Foil Gauges

Open faced gauges must have their grids protected from flux splashes and accidental tinning. Drafting tape is a very suitable material for this task as its mastic is reasonably gentle and easily removed with solvent when soldering is complete, but at the same time it is able to withstand the heat from the soldering iron. The grid masking operation also allows the operator to control the position of the joint on the gauge tab as it is not normally necessary or desirable to use the whole tab area. The maximum fatigue life of a foil gauge will be more easily attained if the joint is kept as small as possible and towards the back of the tab area as far away from the grid-to-tab junction as possible. The masking method also ensures joints of equal size are formed on adjacent tabs thus providing more effective cancellation of any thermal EMFs. If the gauge has been installed and overcoated with adhesive or provided with a protective overlay by the manufacturer the masking step can be omitted. Following masking, the exposed tabs should be cleaned of any mastic residue using solvent and either a cotton bud or small brush. They are then tinned by placing cored solder directly on the tab, and placing the iron on top of the solder using moderate to firm pressure.

A simple rule for the beginner is to count 1000, 2000 (i.e. about 2 sec) whilst keeping the iron in contact with the tab and at the same time feeding in solder to maintain a supply of fresh flux at the joint, followed by immediate removal of solder and iron together. With a little practice and correct iron temperature a smooth hemispherical joint will be formed which is both easy to clean when removing flux and free of spikes that may penetrate the protective coating during use. If the tab does not readily tin, a glass-like layer of flux residue will form after repeated attempts, preventing tinning. This residue should be removed using solvent. Badly oxidised tabs or terminal strips may be cleaned before tinning by lightly lapping the exposed area using a slightly moistened cotton bud and either pumice powder or fine aluminium oxide. Lapping should be accomplished by using single strokes applied down the tab working away from the grid which must be protected with masking tape, all residues being removed with solvent. Before attaching the lead wire it is a good plan to remove any flux residue, the leads selected should then be tinned and positioned on the gauge tab or terminal strip and held in the required place using a small piece of masking tape. If the wire has a thick insulation which keeps the conductor off the surface, the conductor should be bent or 'joggled' so that it is in contact with the foil, to give a thinner solder joint and avoid any 'spring' in the wire which would delaminate the foil. Using masking tape to hold the wire in place leaves both hands free to handle the solder and the iron using the same technique as described for tinning. The result should be a smooth low profile joint from which flux is easily removed (Fig. 12).

FIG. 12. A standard stress analysis installation showing the use of flat three conductor cable for quarter bridge operation. Note the use of a single strand of the conductor to form the jumper wires. Also note the strain relief loops and the gap between gauge backing and the terminal strip. Courtesy of Micro Measurements.

Solder Dots

Some types of strain gauges are supplied with presoldered tabs in the form of small precisely located solder dots.

Whilst these dots are a useful option the correct soldering technique must be used to ensure a sound joint.

If the strain gauge is of the open faced type it should be coated with adhesive at the time of installation. This is a necessary step to prevent the solder dot from spreading during the tinning and lead wire attachment operations. The presence of a cured layer of adhesive over the grid also means that masking of the grid as for a normal open faced gauge is not required.

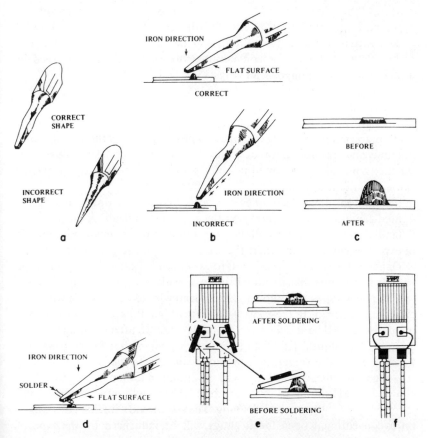

FIG. 13.   Soldering technique for lead attachment to strain gauges with solder dots. Courtesy of Micro Measurements.

Coating of the grid with adhesive is not required if the gauge is of the pre-encapsulated type. In this form of construction access to the solder dot is provided by cut-outs in the encapsulation.

The initial operation is to remove any adhesive covering the solder dot. This can be done by carefully scraping the surface of the dot with a scalpel or by lightly abrading with silicone carbide paper.

The second step is to perform a tinning operation. This is a most important step and no attempt to attach lead wires to solder dots should be made without first tinning the dot to produce a generous hemispherical shaped bead of solder (Figs 13(b), (c) and (d)).

To avoid the possibility of sucking all the solder out of the dot area the soldering iron tip temperature should only be high enough to ensure good wetting of the solder. The iron tip should also be of the chisel or screwdriver type and in no circumstances should it have a sharp point (Fig. 13(a)).

The actual operation of tinning can be carried out using the same technique as described for conventional open faced gauges followed by attachment of the jumper lead (Figs 13(e) and (f)).[14]

Pre-Attaching Lead Wires to Unbonded Gauge

Plastics and some modern composite materials can present problems when soldering as many are easily damaged by heat and also by the solvents used to remove flux residue. Great care must be taken to select a suitable solvent for flux removal. A low melting point solder helps reduce the problems caused by the heat from the soldering iron.

Installations down deep holes or recesses where soldering iron access is limited can also make soldering on the gauge tag difficult. A third area of difficulty arises when an installation must be carried out in an inflammable or explosive environment where the use of electrical equipment is restricted.

One solution to these problems is to use preleaded gauge's or a gauge type to which it is possible to attach jumper leads before bonding.[14] Some special gauge types also allow the main lead wire to be attached while the gauge is unbonded. In these cases, for ease of handling, only short lengths of main lead need be attached and the cable extended after bonding by using crimping techniques. This is a particularly useful method in environments where the use of a soldering iron is difficult or restricted.

Gauges with flexible backings and pre-attached terminals make the best choice for preleading as these are best able to withstand the handling involved when soldering and cleaning. The size of lead wire that can be used on a conventional open faced gauge will be influenced by the type of adhesive to be used. Generally a diameter up to 0·8 mm (including insulation) can be used when using 100 % solids epoxy but this should be reduced to 0·25 mm if a solvent-thinned adhesive is to be used because a more uniform clamping pressure is normally required.

The following is a summary of the stages involved in preleading a conventional open faced gauge:

(1) Using a *clean* steel or aluminium mounting plate (to provide a good heat sink), mount the gauge using masking tape to both protect the grid and ensure good thermal contact between gauge and plate. Leave only the desired amount of tag exposed.

(2) Apply a small amount of flux to the tab *or* to the prepared lead wire (lead wire should have been previously tinned).

(3) Place lead in position on gauge tab and hold down with masking tape.

(4) Tin soldering iron bit with a small amount of solder (chisel or screwdriver shape preferred.)

(5) Apply iron to lead wire tab assembly and hold in position for one second.

(6) Thoroughly wash gauge assembly with flux removing solvent, sufficient to 'float' away the restraining tapes.

(7) Scrub gauge assembly with fine brush while submerged in a container of solvent.

(8) Blot gauge dry between layers of gauze sponges.

(9) Using a cotton applicator slightly moistened with neutraliser wipe the bonding face of the backing.

(10) Place into a clean container.

Silver Soldering

This method must be used when operating above approximately 290 °C and normally requires short lengths of lead wires attached to the strain gauge by the manufacturer. No terminal strip is used but it is necessary to provide an insulated area on the surface immediately behind the gauge by bonding a piece of polyimide or glass cloth tape during gauge installation. A typical lead wire suitable for this temperature range consists of solid nickel-clad copper conductor with fibreglass insulation. The joint is formed by using a solder paste incorporating a flux and a resistance soldering tool to provide the required heat. The resistance soldering tool consists of a set of hand operated tweezers incorporating the electrodes connected to a control unit, to vary the current to the electrodes, and a foot operated on–off switch to leave both hands free to manipulate the strain gauge lead wire. Approximately 1 cm of fibreglass insulation should be removed from the wire, having first flame burnished about 25 mm to prevent fraying. The bared section should be formed into a right angle bend starting close to the insulation. Forming snake like bends in the round conductor on the surface of the specimen helps prevent the lead wire rolling about while performing the silver soldering operation. With the lead wire positioned behind the strain gauge, on the insulated area, and restrained by a piece of tape, the gauge leads are wrapped tightly around the section of bared conductor close to the corner of the right angle bend. A small amount of silver solder paste is placed on the wrapped joint followed by gripping the bared conductor near its end with the electrodes. Operating the foot switch passes a current through the wrapped joint causing it to heat up and melt the silver solder thus forming the joint. The tweezer grip must be released before switching off the current to prevent the tweezer electrodes remaining attached to the lead wire. When cool the joint will be covered with a hard glass-like flux residue which *must* be removed using metal conditioner and neutraliser.[9]

Welding
Above 650 °C spot welding is the only method available and is generally used for attaching leads to strippable backed gauges such as would be used in conjunction with ceramic or flame sprayed alumina. Some conventional organically backed gauges are also available with leads suitable for spot welding which may in some circumstances be preferred to soldering. In these cases, if a round conductor is supplied, a more satisfactory joint will be obtained if the end of the conductor is 'flattened' before welding. In the case of strippable backed foil gauges the leads normally take the form of ribbons of nickel–chrome approximately 0·001″ thick and 0·062″ wide. Using short lengths of about 1″ welded to the gauge tab prior to installation makes handling easier and these can be extended after installation of the gauge.[10]

PROTECTION

Protection is almost the final stage of the installation process with the result that there is sometimes a temptation to cut corners when under pressure of time. It is also frequently a stage to which inadequate thought is given at the planning stage with the result that otherwise sound installations prove unsatisfactory or fail much earlier than anticipated when put into service.

All strain gauges should be protected, including those that are only going to be used in air at room temperature, because even touching an unprotected grid with a finger can lead to a permanent change in the gauge resistance which will be seen as a shift in the zero datum. Protection from mechanical damage as well as the effects of corrosion and oxidation must therefore be provided. In addition to the grid, the gauge backing and adhesive layer require protection to prevent dimensional instability of the backing due to swelling and possible irreversible damage to the adhesive layer. Damage to the grid, corrosion, and the effects of oxidation are usually seen as a permanent increase in gauge resistance, whilst a drop in the value of insulation resistance is seen as a decrease in gauge resistance.

The degree of protection required may vary from a simple single component system for short-term room temperature operation in laboratory conditions, to more elaborate and sophisticated multilayer, multicomponent systems for the more hostile environment such as long-term exposure to sea water, hydraulic fluid and high temperature. Long-term use in these more extreme environments may involve the use of metallic covers incorporating metal conduits for the lead wires and the use of the special techniques such as hermetic sealing or purging with an inert gas. The protection of gauges in hostile environments is the subject of a separate chapter.

A wide range of materials is available from many sources capable of providing suitable protection against the various environmental problems likely to be encountered and it should be relatively simple to select a material able to withstand a particular environment for short-term protection, but it is important to ensure that the material will have no harmful effects on the gauge grid, backing or adhesive. If the material is to be used as part of a multicoating system it must also be established that all the materials to be used are compatible with each other, including the lead wires.

In selecting an appropriate protective coating system consideration must be given to any reinforcing effects that might be produced. If reinforcing to some degree cannot be avoided and calibration of the installation is not possible, due account of the effect on the results must be made and some experimental work may be necessary. Changes of temperature may lead to changes in the modulus of elasticity of the coating system leading to a change in the reinforcing effect. The effectiveness and life of a coating system can be directly related to the care taken during its application. Good surface preparation is essential and the same principles apply as previously described for bonding a gauge, although in practice it is usually difficult to achieve with a gauge and associated lead wires already in place. Whatever preparation technique is used the strain gauge and terminal area should be protected during the preparation process. This can be done by covering the gauge and terminal with masking tape. The following check list should help to maximise coating life.

(1) Carry out best possible surface preparation.

(2) Ensure that prepared area is sufficient to allow successive coating layers to overlap.

(3) Remove all flux.

(4) Pretreat lead wires as appropriate.

(5) If possible track lead wire around in coating area to achieve longest possible path between gauge and lead wire exit from coating.

(6) Lift lead wire away from the surface as soon as possible after gauge or terminal strip connection and 'bury' in coating material.

(7) Lead wires should be separated into single conductors before exit from the coating to allow for proper pretreatment and application of the coating material round the whole of each conductor.

(8) Carry coating along lead wire for a short distance (i.e. 2 to 3").

(9) Arrange for a method of anchoring lead wires outside coating to minimise movement of the lead wire where it leaves the coating.

(10) Ensure complete curing of each 'layer' of coating material before applying a further layer.

(11) Do not 'feather' the edges of the coating and avoid 'sharp' corners around the boundary.[11]

VERIFICATION

Before putting a strain gauge into service the quality of the installation should be verified by checking the electrical properties of leakage-to-ground, (insulation resistance) and installed resistance, together with some form of check on the bond. While the first two checks do not offer conclusive proof that the installation is satisfactory they do give a good indication of possible trouble. Testing should not be confined to the completed installation but carried out at various stages during the installation process to identify faults at an early stage thus allowing easy rectification. The minimum number of stage checks should be: after bonding; after attachment of the leads; and after application of the protective coating system (in the case of the latter, after each coating layer).

Checking the installed resistance should show up any damage to the grid caused by poor handling, the clamping system, or the effects of such things as flux residue. In the case of a conventional open faced foil strain gauge with a room temperature curing adhesive the installed resistance should be close to or within the manufacturers tolerance, however 'hot' cured installations and gauges installed on small radii may exhibit larger variations.

The leakage-to-ground check should help identify problems such as damage to the backing, flux residue, trapped moisture, solvents, or foreign matter, and indicate incomplete curing on the adhesive or protective coating. A well installed gauge should exhibit an insulation resistance in excess of 20 000 megohms, although this value may drop a little after application of the protective coating depending on the material used. An acceptable minimum value is 10 000 megohms. In tests where the temperature exceeds 150 °C lower values will be experienced and in this case 10 megohms should be the minimum value accepted. As part of a well planned installation provision should be made to record at least the final insulation and installed resistance values as these will be found an invaluable aid in diagnosing the cause of any problems that might arise during the life of the installation. Verification of the bond is a more difficult problem and most satisfactorily carried out by a full scale function check of the installation by applying the expected loads or deflections and observing the installations behaviour and its ability to return to the original zero. Unfortunately a full function check is frequently not possible due to the nature of the test or the size of the structure.

In such cases a useful, although not conclusive, test that may be used for foil or wire gauges is to lightly press on the grid using a soft pencil eraser. In the case of a sound bond, free of voids or bubbles, the gauge's zero will shift slightly as pressure is applied but immediately return to the original zero when pressure is removed.

REFERENCES

1. STEIN, P. K., 1960, Adhesives, how they determine and limit strain gauge performance, Stein Engineering Services, Arizona, USA.
2. *Surface Preparation for Strain Gauge Bonding*, 1976, B-129, Measurements Group, North Carolina, USA.

3. *Strain Gauge Installations with M-Bond 43B, 600 and 610 Adhesive*, 1978, B-130-6, Measurements Group, North Carolina, USA.
4. *Strain Gauge Installations with M-Bond 200*, 1976, B-127-5, Measurements Group, North Carolina, USA.
5. *The Proper Use of Bondable Terminals in Strain Gauge Applications*, 1975, TT-127-4, Measurements Group, North Carolina, USA.
6. VORNEHM, R. C., Unique Clamping Techniques for strain gauges, 1980, *Experimental Techniques Volume 4*, No. 5.
7. *Noise Control in Strain Gauge Measurements*, 1980, TN-501, Measurements Group, North Carolina, USA.
8. STEIN, P. K., The response of transducers to their environment, The problem of signal and noise, 1969, *LF/MSE Publication No. 17*.
9. *Silver Soldering Techniques for Attachment of Leads to Strain Gauges*, 1976, TT-129-4, Measurements Group, North Carolina, USA.
10. *Strain Gauge Installations with M-Bond GA 100*, 1978, B-132, Measurements Group, North Carolina, USA.
11. POPLE, J., *BSSM Strain Measurement Reference Book*, 1980.
12. *Attachment Techniques for Weldable Strain Gauges and Temperature Sensors*, 1977, B-131-2, Measurements Group, North Carolina, USA.
13. WEIDNER, J., *Transducer Installation Techniques*, unpublished.
14. *Techniques for Attaching Leadwires to Unbonded Strain Gauges*, 1978, TT-134, Measurements Group, North Carolina, USA.

APPENDIX I

Installation of Weldable Strain Gauges

As their name implies weldable strain gauges do not require an adhesive to attach them to a structure but are spot-welded into position using a capacitive discharge spot-welder with a 0–40 watts/second weld energy range and spherical electrode tip of typically 0·8 mm diameter. They are limited to use on ferrous materials. Weldable gauges are commercially available in several forms, the cheapest and simplest type, for temperature ranges up to 260 °C, consisting of a conventional foil strain gauge prebonded to a stainless steel carrier approximately 0·005″ thick. The gauges normally require the attachment of lead wires and protection in the same manner as a conventional gauge.

A second type for use to 600 °C uses a special wire resistance element in quarter or half bridge form sealed into a stainless steel tube which is in turn spot-welded to a 0·005″ stainless steel shim. These gauges are normally supplied with leads and offer a high degree of environmental protection. A further form of construction employs wire grids in full bridge form prebonded to a steel carrier with ceramic cement.

In all cases surface preparation requirements are minimal and therefore weldable gauges offer advantages when an installation must be carried out quickly or when the environmental conditions would make it difficult to use conventionally bonded gauges.

Surface Preparation
The gauge location should be free of grease, rust, scale oxides and surface irregularities. This can generally be accomplished in three stages.

(1)   Solvent degrease the area.
(2)   Hand grind, disc sand or file the area smooth.
(3)   Wash area with solvent to remove residues.

After stage (2) the surface should be free of any significant pitting or grooves that might cause a flash back during welding leading to possible damage to the carrier.

Gauge Preparation and Attachment
Normally no preparation is necessary apart from attaching a short length of drafting tape across the width of the gauge to provide a means of

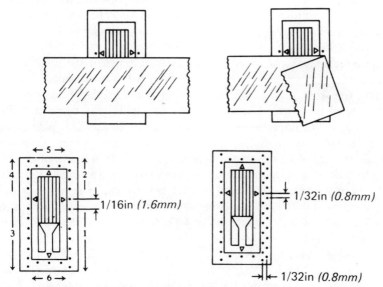

FIG. 14.   Welding sequence for weldable strain gauges. Step 1: Drafting tape holding gauge in position on specimen prior to tack welding. Step 2: Removal of tape after tack welding gauge in position. Step 3: First row of welds complete. Step 4: Spacing of second row of welds. Courtesy of Micro Measurements.

positioning the gauge on the structure prior to welding. Conventional marking out techniques can be used if required.

It is important to follow the manufacturers recommended welding sequence (Fig. 14) for a particular type of gauge, also a practice weld using a piece of carrier material should be performed on the test structure to determine the proper weld energy setting. With a satisfactory weld a small slug of metal should pull away from either the test surface or the carrier when the latter is pulled away.[12]

APPENDIX II

Surface Preparation—Grit Blasting Method

When a large number of gauge areas need to be prepared 'grit blasting' may be considered as an alternative to the universal method. The technique produces a more even finish at a much faster rate than is possible with other methods but uses a certain amount of normally fixed equipment in the form of a compressor, dust extraction unit and work chamber and is therefore not generally suited to on-site work. It is widely used in the transducer industry. As described for the universal method it may be necessary to carry out some preliminary preparation before the grit blasting as the technique can very effectively mask an unsuitable surface.

If the technique is to be used attention should be paid to the quality of the compressed air used and the handling of the grit. The air supply must be filtered free from oil and water contamination and regular checks made on the state of the filters. Use of a works air supply may be possible but suitability of the filtering arrangements must be established. Where possible a separate air supply to be used only for the grit blasting equipment should be used. Air pressures of between 75–80 psi (5–6 bar) are normally required, depending on the equipment used, producing a nozzle velocity of the abrasive stream of around 1000 ft per second. For some critical applications an alternative to compressed air can be used in the form of bottled inert gas such as nitrogen or CO_2 (note: oxygen should *never* be used due to its flammability).

A recommended abrasive powder is aluminium oxide with a size of approximately 29 microns producing a surface finish of 20 microinches r.m.s (0·5 micrometres r.m.s.) A 20 microinch finish is ideal for transducer work and many stress analysis applications but would normally be too fine for high elongation or ceramic cement applications where an abrasive capable of producing a coarser finish of around 250 microinches r.m.s. (6·4 micrometers r.m.s.) should be used. Spent abrasive powder should *never* be recycled as it may be contaminated, in addition the powder loses its cutting points becoming less efficient.

The complete surface preparation cycle using grit blasting is a three stage process.

(1) Solvent degreasing.
(2) Abrading.
(3) Ultrasonic cleaning (post abrading).

Solvent Degreasing
The first stage of solvent degreasing should be carried out as previously described for the universal surface preparation method and the same considerations prevail.

Abrading
Using the hand tool provided with the grit blasting equipment abrasion is accomplished by using even strokes, similar to a spray painting action, keeping the nozzle 2–3 in away from the surface and at an angle of 45–60°. The hand tool attached to the powder delivery hose is typically the size and shape of a ballpoint pen and is fitted with a nozzle incorporating an orifice of approximately 0·046 in across the flats.

Ultrasonic Cleaning
Following abrasion the component should be placed in an ultrasonic solvent bath for final degreasing and removal of any abrasive grit still on the component surface.

If an ultrasonic tank was used for the initial degreasing the same tank should *not* be used for the post abrading cleaning as it is particularly important that contamination should not be allowed to build-up in the tank used for the final cleaning step. Solvents such as Freon, isopropyl alcohol or Arklon are normally employed and changed frequently to minimise the build-up of contamination.

Gauge Location
The same considerations apply as discussed previously. If marking on the component is required this should be carried out with an inkless ballpoint or hard pencil to produce a burnished line. The surface should then be scrubbed with neutraliser and cotton swabs to remove any contamination caused by the operation. As with the universal method of surface

preparation a cleaned surface will oxidise in the presence of air making it important that bonding takes place as soon as possible after preparation. The time lapse that may be permitted varies depending on the material but typically titanium should not be left more than 10 min, aluminium, beryllium copper and brass 30 min, and steel not more than 45 min before bonding.

Toxic Materials

Beryllium, beryllium copper and lead along with other toxic materials require special handling and should not be abraded using the standard grit blasting method.

Expert safety department advice should be sought on methods of material removal. Debris from the abrasion process must not be allowed to become airborne and should be kept wet.[13]

APPENDIX III

Porous Surfaces (Cast Iron and Concrete)

Rough or porous surfaces require a slightly modified installation technique. This is an important class of materials typified by cast iron and concrete, two materials to which strain gauges are frequently applied. This appendix suggests methods of dealing with both materials. The same technique can be used for other materials of similar nature.

The principal requirement is to provide a smooth surface completely free of loose material and voids in the gauge area. In any installation voids under or around a gauge grid degrade strain transmission into the grid and give rise to stability problems once the gauge is energised due to the lack of heat sink in the void area. Most serious problems arise if the installation is also subject to hydrostatic pressure. Large zero offsets will occur as pressure is applied, caused by the grid deforming into the voids. The offset will be largely non-recoverable when the pressure is removed. In the case of a well installed gauge these offsets will be small, in the order of 10 microstrain per 1000 psi and recoverable when pressure is removed.

The basic technique when dealing with porous surfaces is to fill the voids and irregularities with a precoat of suitable adhesive after a modified surface preparation. After the precoat has been cured it should be lapped with a suitable grade of silicon carbide paper until the porous substrate begins to show through the adhesive layer. The surface should then be carefully examined and if voids are still present the filling operation repeated. Once the surface is satisfactory a standard installation can be carried out. Cyanoacrylates and low viscosity adhesives such as solvent-thinned types are not suitable for the precoat stage. Reinforcement of the surface is not normally a problem.[2]

Surface Preparation
Cast Iron
A major problem with cast iron is the difficulty of achieving effective degreasing. If circumstances permit, the component should be heated for 2–6 h at 175 °C to drive off volatile hydrocarbons and break down lubricating oils etc. followed by alternating degreasing and heating until the surface is clean. On large or fixed structures a propane torch may be used to achieve this. The surface may then be disc sanded or filed to remove surface roughness followed by wet abrading with conditioner, marking of gauge location lines, and neutralising as described under the universal surface preparation method. The surface must then be completely dried before application of the adhesive precoat.

Concrete
The use of concrete for civil engineering projects means the surface is often heavily contaminated with soil and other debris from the building operation. This contamination can be removed by using a stiff bristled brush and detergent, followed by thorough rinsing with water. Less contaminated surfaces and light surface grease can be simply removed by a wipe through the gauge area with a gauze sponge soaked in a volatile solvent, the operation being repeated until all contamination is removed. Concrete surfaces may require irregularities to be removed by use of a sanding disc followed by brushing with a wire brush. The resulting debris should be removed from the surface by using clean compressed air when available or by scrubbing with clean water. The surface should then be flooded with conditioner and scrubbed with a stiff brush followed by blotting to remove contaminated conditioner and a thorough wash and rinse with clean water. A liberal quantity of neutraliser should be applied to the surface to restore the surface alkalinity to a pH of approximately 7, blotted, and the surface finally washed and rinsed with clean water. Before the necessary adhesive precoat is applied the surface must be thoroughly dried, using a heat lamp, hot air blower or propane torch if necessary.[2]

Chapter 3

GAUGE INSTALLATION AND PROTECTION IN HOSTILE ENVIRONMENTS

J. POPLE

*Vickers Shipbuilding and Engineering Ltd,
Barrow-in-Furness, UK*

INTRODUCTION

In terms of strain measurement, an environment may be defined as hostile if it has a detrimental effect on the measurement system: experience has shown that any deviation from 'ideal laboratory' conditions is potentially hostile. Certain environmental aspects may not be at all obvious, nor indeed reasonably foreseeable, for example, the unwelcome attentions of the idly curious, though with forethought even these effects can be minimised.

The choice of strain gauge materials and techniques is determined by the expected environmental conditions, and when making an assessment of these, the installation phase and the premeasurement and measurement phases must be equally considered. The measurement environment is usually easier to define in terms of temperature, fluid immersion, strain level etc. but here the most important consideration is often the time factor, i.e. survival time at temperature or duration of immersion. The installation and premeasurement environments can prove to be less easy to define in advance, and may well be the limiting factors, though this may not be immediately apparent: a quick look at a drawing, for example, may fail to indicate that whilst conventional foil techniques are quite satisfactory from an operational viewpoint their application is made impractical, because of limited access to the gauging area.

In addition, a strain measurement requirement may be initiated by someone unfamiliar with the practicalities of such a measurement, and hence environmental aspects may well be ignored or overlooked; therefore,

85

even the most seemingly straightforward measurement tasks should be thoroughly researched by the installation team. This chapter deals specifically with the application of electrical resistance strain (ERS) gauges, and where environmental considerations make demands beyond the practical limits of these devices other measurement systems must be employed (see Part II Other Strain Gauges).

To illustrate the processes involved in the formulation of a strain gauge installation, two commonly encountered environmental constraints will be considered, namely temperature and fluid immersion.

The first case, temperature, is perhaps the most frequently encountered problem and as it is not normally possible to protect an installation from the effects of temperature, success is dependent on the correct selection of components and materials for survival over the range of temperatures envisaged. In the second case, fluid immersion, it may well be possible to protect a conventional installation from the effects of the environment by the employment of suitable materials and techniques. The two major factors involved are, therefore, selection for and protection from the particular environmental conditions under consideration.

Temperature has probably the most detrimental effect on a strain gauge installation, and considerable problems must be expected where measurements are required over a wide range of temperatures or where materials are used at the limits of their specified temperature range. In the latter case, the limiting factor could well be the gauge/adhesive combination itself and, unfortunately, the fact that ERS gauges are normally easier to apply and are cheaper than other high temperature strain measurement devices can result in their being used at or above the manufacturers' recommended temperature limit. The problem is compounded by the lack of information on the survival time at temperature of gauge/adhesive combinations, which can lead to unexpected failure caused by breakdown of the adhesive; break-up of the gauge backing material and oxidisation effects on the gauge foil, solder joints, cable and connectors. Deterioration is very much a function of time-at-temperature and can therefore preclude any long-term measurement capability. Temperature is emphasised because its effects are a major cause of non-strain induced resistance changes, which result in an unreliable strain measurement system, and because the problems involved aptly illustrate the first criterion for successful measurements in a hostile environment—the selection of materials for survival. The second criterion, protection of the gauge installation from the hostile environment, often receives little consideration and, consequently, can be a major source of failure.

Detailed examination of all the hostile environments in which strain measurements with ERS gauges are practicable is beyond the scope of this chapter but some of the most commonly experienced are listed in Table 1. A bibliography of specialist papers selected from reference 1 is given at the end of the chapter. Before attempting strain measurements in any hostile environment, the user is strongly advised to consult all available up-to-date literature, to ascertain current practice. Many aspects of accepted gauging

TABLE 1
STRAIN MEASUREMENT HOSTILE ENVIRONMENTS

Hostile environment	Bibliographical reference
Low temperature	1–3
High temperature	4–6
Transient temperature	7, 8
Fluid immersion	9–11
Fluid pressure	12–16
Fluid flow	17–19
Nuclear radiation field	20–24
High gravitational field	25
Electrostatic/electromagnetic field	26, 27
Vacuum	1, 20
Blast	28
Hazardous atmosphere	29, 30
Embedded	31, 32
Rotating shafts	33
Fatigue	34
High elongation	35, 36
Non-metallic materials	37, 38
General	20, 22, 39, 40

practice, and much generally available manufacturers' literature, relate to an 'ideal' or 'laboratory type' environment, and whilst much of this may be applicable in a given hostile situation, it should not be applied without careful consideration—in some situations a quite markedly different approach and a whole new way of thinking will need to be adopted.

With the foregoing in mind, the four main aspects of strain measurement in a hostile environment will be considered:

(a)   choice of materials;
(b)   installation techniques;
(c)   protection;
(d)   obtaining reliable data.

CHOICE OF MATERIALS

Before starting the process of selection of materials for a particular measurement exercise, certain factors must be clearly defined in addition to the environmental conditions. Figure 1 illustrates the eight major inter-related factors and Table 2 lists the main points for each. It is of paramount importance to obtain information on all these main points before starting the material selection process.

One consideration which is often overlooked is survival of the installation during the premeasurement period. An example from the author's experience will amply illustrate the problems. Measurements were required to determine the effects of pressure and gauges were laid on the inside and outside of a structure. Testing involved filling the structure with water and applying pressure, over a period of less than 24 h. Considering the gauges on the outside, the measurement environment was atmospheric conditions, i.e. ambient temperature and humidity. However, examination of the gauge installation requirements indicated that, owing to the nature of the fabrication process the gauges inside had to be installed 12 months prior to the test, and subsequently became inaccessible. During this 12 month period, the structure was subjected to a heavy machining operation within 3 cm of the gauges, using cutting fluids capable of reducing a gauge to its basic components (i.e. foil, backing etc.) in less than 5 h and of destroying the adhesive bond within 3 h. In addition to this, a 'weld pass' temperature in excess of 250 °C for 5 mins could be expected. So, what at first sight appeared to be an 'ideal laboratory type' installation was, in fact,

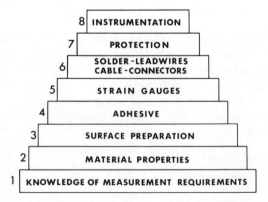

FIG. 1. Metal foil gauges installation considerations.

TABLE 2

TABLE 2

METAL FOIL GAUGES INSTALLATION CONSIDERATIONS

	Main points
1 Knowledge of measurement requirements	(a) correct measurement technique (b) environment—both installation and test (c) test duration (d) measurement accuracy (e) dynamic or static (f) temperature range—both installation and test (g) type of stress field and strain gradient
2 Material properties	Knowledge of: (a) Young's modulus (E) (b) Poisson's ratio (μ) (c) coefficient of thermal expansion (α) (d) construction, e.g. laminated
3 Surface preparation	Availability of: (a) preparation time (b) mechanical aids, also limitations on use of chemicals and amount of material removable
4 Adhesive	(a) temperature—installing and test (b) strain levels (c) fatigue life (d) elongation capabilities (e) compatibility with gauge backing and material (f) clamping availability and accessibility
5 Strain gauges	Compatibility with: · (a) adhesive (b) strain levels (c) fatigue characteristic (d) temperature and environment (e) instrumentation (f) material constants (g) correct geometry and site for stress field
6 Solder, lead wires, cable connectors	(a) bridge configuration (b) gauge factor desensitisation (c) corrosion (d) temperature (e) mechanical strength (f) shielding
7 Protection	Survival in environment having: (a) temperature range (b) mechanical abrasion (c) chemical attack (d) pressure (e) able to cure in installation environment
8 Instrumentation	(a) bridge excitation—type, mode and level (b) errors (c) digital or analogue (d) transient and dynamic response (e) calibration—method and traceability

one in which survival of the installation prior to the measurement phase was the dominant concern.

A key factor in the survivability of a strain measurement system is the choice of suitable materials. Examination of failed installations has indicated that a high proportion of failures are attributable to the use of unsuitable materials. A series of laboratory experiments may be necessary to realistically simulate the installation environment and to determine the effects of that environment on the materials and techniques being considered. (Full details on how to set up a suitable test programme are given later.) Experience has shown that when test data obtained in the laboratory is to be related to a hostile environment, the relevance is greater when the data has been obtained using high quality strain gauge materials. Indeed, it is the author's firmly held opinion that in all hostile environment situations, the only cost-effective strain gauge materials are those produced by a manufacturer who can demonstrate the use of a quality control system. The cost failure of the complete installation needs to be compared with the marginal savings obtained from using cheap or inferior materials. Selection of strain gauge materials should be governed by such factors as the accuracy required, survival in the hostile environment and the reliability expected from the strain measurement system, rather than by cost alone.

With all these aspects in mind, consideration will now be given to the environmentally-dependent aspects of selection of:

    (a)   gauge foil and backing;
    (b)   adhesives;
    (c)   terminals;
    (d)   cable;
    (e)   solder;
    (f)   plugs, sockets and connectors.

Gauge Foil and Backing
Developments in the manufacture of ERS strain gauges have produced a foil gauge based on the characteristics given in Table 3.

A range of high quality gauges is available from which the user must select a particular gauge to suit his requirements. The five main areas of consideration are given in Table 4 and it will be noted that many of the points are environmentally-dependent. Insufficient attention to these points will result in failure of the measurement system due to the adverse effects of a hostile environment.

TABLE 3
FOIL STRAIN GAUGE CHARACTERISTICS

Backing	Strain transmission
	Flexibility
	Stability
	Elongation properties
Foil	Resistance stability with time and temperature
	Elongation linearity
	Fatigue life
	Gauge factor and Kt in uniaxial stress field with known Poisson's ratio
	'Thermal output'
Types	Single, double and three-grid configurations
	Combinations of backing and foils (often termed gauge series)

Three major considerations are evident, namely the limits of temperature, strain level and fatigue life. Some manufacturers, quite rightly, quote different upper temperature limits for static and dynamic measurement applications but few give survival-time-at-temperature data.

The most commonly used gauge has a constantan foil with a polyimide backing material and has a quoted temperature range of, typically, $-200\,°C$ to $+250\,°C/+300\,°C$. However, the useful life at the upper temperature is rated in hours rather than days. The standard constantan foil gauge is available with a self-temperature-compensation (STC) characteristic and a strain range of up to $\pm 5\%$. In a 'super-annealed' form this gauge type has an extended strain range of typically $\pm 20\%$ and although the temperature range is reduced in consequence to, usually, $0\,°C$ to $100\,°C$, this can be extended to $-75\,°C$ to $+200\,°C$, for short duration measurements, by employing a full bridge configuration. Typical quoted fatigue life limits are ± 1500 microstrain for 10^6 cycles for the standard form, and ± 1000 microstrain for 10^4 cycles for the super-annealed form.

Different foil alloys are employed to produce a gauge with improved time-at-temperature characteristics and to slightly increase the operating temperature range ($-270\,°C$ to $+300\,°C$). Of these, the most commonly used is a modified Karma alloy (Ni, Cr, Fe, Cu) and although STC gauges are available, it must be noted that at the upper temperature limits the STC characteristic may be destroyed, and that permanent changes in resistance can occur with time. This gauge foil has a better fatigue life characteristic than constantan, being typically ± 2200 microstrain for 10^7 cycles.

Above $250\,°C-300\,°C$ the choice of ERS gauges is considerably restricted with respect to configuration, fatigue life and strain range. A

TABLE 4
FOIL STRAIN GAUGE SELECTION CONSIDERATIONS

Backing material	Operating temperature, both long-term and short-term. Compatibility with selected adhesive.
Gauge foil	Strain and temperature range. Fatigue life. Self-temperature-compensation (STC).
Grid geometry	Single-grid—uniaxial field with known principal axis. Two element grid—biaxial field with known principal axes. Torque measurements. Three element grid—where the principal axes are unknown or the maximum and minimum values need to be determined.
Grid size	Grid size affects bridge voltage and output levels. Small grid for high strain gradients (e.g. active gauge length 0·032 in.) Large grid for average strain measurement over an area. Most commonly used active grid length—0·125 in.

Grid resistance	Advantages	Disadvantages
120 ohm	Less shunting effect of insulation breakdown. Produces a purer constant voltage source.	Higher bridge voltages and output signals possible. Less effect of lead resistance on gauge factor desensitisation.
350 ohm	Only lower bridge voltages and output signals obtainable. More effect of lead resistance on gauge factor desensitisation.	More shunting effect of insulation breakdown.

platinum–tungsten alloy gauge is available for operation in the temperature range $-200\,°C$ to $+650\,°C$ but this takes the form of a strippable backing, open-faced free grid, which must be used with special ceramic adhesives, presenting handling problems which could prove a limiting factor in hostile installation environment applications. These gauges are not available in STC form, and alternative compensation techniques are necessary, particularly where low strain levels and large temperature changes are expected. The strain range is limited, becoming non-linear above $0·3\%$. Should the platinum–tungsten gauge prove unsuitable, consideration should be given to the use of those more sophisticated, and more expensive techniques specifically developed for high temperature strain measurement (see Part II Other Strain Gauges).

To simplify the gauge selection process, and to ensure that all the relevant points have been considered, a selection 'check list' (based on reference 2) is an invaluable aid (see Table 5). The choice of which gauge to use can range from the simple 'what is in the cupboard?', to ploughing through the manufacturers' catalogues (one manufacturer lists over 40 000 different gauges), choosing the 'ideal gauge' and then waiting for delivery. A cautionary note regarding the 'what is in the cupboard?' approach: when a

TABLE 5
FOIL STRAIN GAUGE SELECTION CHECK LIST

	Considerations for parameter selection
Selection step: 1 Parameter : gauge length	strain gradients area of maximum strain accuracy required static strain stability maximum elongation cyclic endurance heat dissipation space for installation ease of installation
Selection step: 2 Parameter : gauge pattern	strain gradients (in-plane and normal to surface) biaxiality of stress heat dissipation space for installation ease of installation gauge resistance availability
Selection step: 3 Parameter : gauge series	type of strain measurement application (static, dynamic, post yield, etc.) operating temperature test duration cyclic endurance accuracy required ease of installation
Selection step: 4 Parameter : options	type of measurement (static, dynamic, post yield, etc.) installation environment—laboratory or field stability requirements soldering sensitivity of substrate (plastic, bone, etc.) space available for installation installation time constraints
Selection step: 5 Parameter : gauge resistance	heat dissipation lead wire desensitisation signal-to-noise ratio
Selection step: 6 Parameter : STC number	test specimen material operating temperature range accuracy required

Courtesy of Micro Measurements.

full bridge configuration is being employed with the intention of achieving temperature compensation, gauges must be of the same type, and from the same batch of foil, to ensure that they share a common STC characteristic. The final choice is usually a compromise—ensuring that all relevant considerations are taken into account.

Adhesives

In the majority of cases, the gauge manufacturer will recommend a suitable adhesive for a particular grid/backing combination for use in a specific application, (e.g. high temperature or elongation). These specialised adhesives have been developed to exhibit the necessary adhesion and strain transmission properties for a satisfactory strain measurement system. Examples are:

(1) good adhesion to the specimen and the gauge backing, under high tensile and compressive strains;
(2) retention of mechanical and chemical stability under a variety of service conditions;
(3) good electrical insulation properties, which are retained under all service conditions;
(4) non-hygroscopic, non-corrosive and compatible with specimen and gauge backing;
(5) not subject to failure as a result of flaws or cracks produced in the adhesive by strains;
(6) amenable to simple bonding techniques.

It should be appreciated that these requirements are so critical that *only recognised batch tested strain gauge adhesives should be employed* and that environmental conditions (installation and measurement) are particularly crucial.

Selection of adhesives must be governed by:

(a) compatibility with the backing of the selected gauge and the test material;
(b) availability of time to cure, ability to realise curing temperature, and adequate clamping facilities;
(c) adequate elongation, operating temperature and 'glue-line' fatigue life characteristics.

Although strain gauge selected cyanoacrylate adhesives are widely used in laboratory environments, they are *not* recommended for use in hostile

environments. Although when properly used they give excellent performance with negligible hysteresis, little creep under load and elongations of
between 6 and 10 %, they are highly sensitive to moisture (greater than 75 %
RH) and have a limited temperature range ($-5°$ to $+65°C$). Another
'ideal laboratory conditions' group of adhesives—epoxies and epoxy–
phenolics—can be used to a limited degree, in hostile environments.
Indeed, because of their upper temperature limit, typically $+250°C$ to
$+300°C$, they are the only strain gauge adhesives available to enable strain
measurements with foil gauges to be made at such temperatures. The
limitations on their use are the adverse effects of a hostile installation
environment upon their application and curing, and difficulties in achieving
effective clamping during curing. The high temperature epoxy and
epoxy–phenolic adhesives require a heat curing cycle which can range from,
typically, 175 °C for 1 h to 75 °C for 4 h at 30–40 psi: limited access to the
gauging area may render this impossible.

One very useful type of strain gauge adhesive for some hostile
installation environments is a polyester. The main advantage is a quick cure
(5–10 mins) at very low temperatures (0–10 °C). These adhesives have an
inferior performance to epoxies but can often enable metal foil gauges to be
bonded in conditions that would otherwise be thought impossible. For
example, with one particular polyester adhesive the author has successfully
installed metal foil gauges in the pouring rain and has developed techniques to
enable foil gauges to be laid underwater.[13] The typical operating temperature range for a cold curing polyester adhesive ranges from $-40°C$ to
$+50°C$ for static measurements, to $+80°C$ for dynamic measurements
with an upper limit of, 120 °C for 1 h. Data on elongation and fatigue
characteristics is not generally quoted, but certain polyester adhesives have
proved capable of allowing the strain level and fatigue life of constantan
foil/polyimide backing gauges to be realised. It should be noted, however,
that polyester adhesives are not compatible with all types of foil gauge
backings, particularly some epoxies. Polyester adhesives also produce a
thicker glue-line than the epoxy–phenolic type.

Full attainment of the properties of a strain gauge adhesive is dependent
to a large extent upon the condition of the uncured adhesive. For example,
whether or not it is 'in date', has absorbed moisture or is otherwise
contaminated etc. These factors are far more critical in hostile environment
applications than in laboratory situations. It must be emphasised that the
manufacturers of high quality gauges quote gauge characteristics obtained
from a sample gauge or batch, using a particular adhesive. When this
particular adhesive is suitable for the user's environment it should be used

in preference to any other, as the quoted characteristics then relate more meaningfully to the measurement exercise. In cases where the hostile environment necessitates the use of a different adhesive, or where gauges from one manufacturer are to be used with adhesive from another, a test on a gauge factor bar should be carried out (see Chapter 5) to demonstrate satisfactory gauge backing/adhesive compatibility. Additional testing will be necessary to verify performance data where high elongation or fatigue life capability is required.

Terminals

In many applications it is advantageous to use a strain gauge terminal to provide the interface between the fine lead wires soldered to the gauge and the heavier cable used to connect the gauge to the rest of the measurement system. The choice to the user is not so varied as that for gauges but nevertheless can be dictated by the environmental conditions. The choice of backing material is temperature related. Two types are commonly available: a Teflon film, which is suitable for long-term use at $+230\,°C$ to $+260\,°C$, limited primarily by the gradual oxidation of the copper foil interface (the relatively high expansion coefficient of unfilled Teflon may cause loss of bond at temperatures below $-75\,°C$) and an epoxy–glass laminate which is strong but quite flexible (this is suitable for long-term use at $+150\,°C$ and has been successfully used in liquid helium at $-269\,°C$).[3] A number of terminal configurations are available for each backing, three of which are illustrated in Fig. 2. The choice of terminal layout is influenced by bridge configuration (two- or three-wire etc.) and the availability of space at the gauge location.[4]

Cable

The correct choice of cable between gauge and bridge completion components is of vital importance. For example, the cable in a $\frac{1}{4}$ bridge configuration forms part of the measurement circuit: the value of the cable resistance and any change in it will, therefore, affect the accuracy of the strain measurement (see Chapter 5).

Cable can be considered in terms of the conductor, the insulation, and the construction. Some commonly used types are listed in Table 6.

In producing standard cables the manufacturer has chosen the conductor/insulation combinations and the user must decide which combination suits the installation and measurement environments; the overriding factor again being temperature. Table 7 indicates typical combinations.

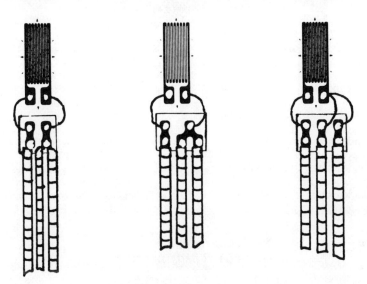

Fig. 2.   Strain gauge terminal configurations.

Other important factors which affect cable selection (some of which are environmentally-dependent), with general points and specific comments are listed in Table 8.

Each particular insulation type has a recommended maximum operating temperature (see Table 9) and in addition to the temperature limitations other hostile environmental restraints must be considered. Table 10 lists a

TABLE 6
COMMONLY USED CABLE MATERIALS

Type of wire	Insulation	Construction
Solid copper	Etched PTFE Teflon[a]	Flat cable
Stranded copper	Fibreglass braid	Twisted cable with jacket
Tinned solid copper	Kapton[a] (polyimide) wrap	Braided shield
Tinned stranded copper	Nylon/polyurethane	Twisted cable without
Silver-plated solid copper	enamel	jacket
Silver-plated stranded	Polyurethane enamel	Round single wire
copper	Polyimide enamel	
Nickel-clad solid copper	ETFE Tefzel[a]	
	Vinyl (PVC)	

[a] DuPont trade name.

TABLE 7
TYPICAL CONDUCTOR/INSULATION COMBINATIONS

Conductor	Insulation	Temperature range
Solid copper	Polyurethane	$-75\,°C$ to $+150\,°C$ ($-100\,°F$ to $+300\,°F$)
Solid copper	Polyimide	$-269\,°C$ to $+315\,°C$ ($-452\,°F$ to $+600\,°F$)
Solid nickel-clad copper	Fibreglass	$-269\,°C$ to $+480\,°C$ ($-452\,°F$ to $+900\,°F$)
Stranded tinned copper	PVC	$-50\,°C$ to $+70\,°C$ ($-60\,°F$ to $+158\,°F$)
Stranded silver-plated copper	PTFE	$-269\,°C$ to $+260\,°C$ ($-452\,°F$ to $+500\,°F$)
Stranded silver-plated copper	Kapton	$-269\,°C$ to $+200\,°C$ to $+300\,°C$ ($-452\,°F$ to $+392\,°F$ to $+570\,°F$)

TABLE 8
FACTORS AFFECTING CABLE SELECTION

Conductor[a]	
Single	Less resistance per unit diameter.
	Suitable for spot-welding.
Stranded	Use in vibration situations.
	Unsuitable for spot-welding.
Plated	Use for high temperatures.
	Difficult to make attachment.
Construction[b]	
Singles	Easier to pressure-test.
	Usually more readily available.
Three-core	Ideal for $\frac{1}{4}$ bridge three-wire with gauge remote from bridge completion.
Ribbon	Produces a neat installation.
	Must be separated *before* entry into coating.
Insulation[c]	
PVC	Tends to become brittle in flow situations.
	Affected by heat and solvents.
	Large variations in quality.
PTFE	Tends to 'cold-flow' at room temperature.
	Requires etching for good bonding.
Kapton	Usually lapped construction.
	Requires etching for good bonding.
ETFE	Good all round insulation.
	Requires no etching for bonding.

[a] Select conductor size to give lowest resistance value possible, but considering cable/terminal size ratio, temperature, vibration and lead attachment method.

[b] Select form to suit protection technique and electrical environment. Use screened cable to reduce pick-up in electrostatic fields and twisted leads to reduce 'loop' area and pick-up in electromagnetic fields.

[c] Select covering to give coating compatibility. Consider temperature, flexibility requirements and cable routeing.

TABLE 9

CABLE INSULATION RECOMMENDED MAXIMUM OPERATING TEMPERATURES

Cable insulation type	Maximum operating temperature ($°C$)	
Polyvinyl chloride (PVC)	70	
PVC heat resisting (PVC HR)	85	
Polyethylene	70	
Rubber	60	
Ethylene propylene rubber (EPR)	90	
Silicone rubber	150	
Tough rubber sheath (TRS)	60	
Polychloroprene (PCP)	70	
Chlorosulphonated polyethylene (CSP)	85	
Nylon	105	
Fluorinated ethylene propylene (FEP)	200	
Polyimide (Kapton)	200–300	
Polytetrafluoroethylene (PTFE)	260	
Tefzel (ETFE)	150	
Insulants		
Glass fibre or tape	200^a	600^b
Asbestos fibre	200^a	500^b
Glass/mica tape	200^a	600^a
Silica fibre	1000^a	1000^b
Impregnants		
Lacquer	90	
Compound	150	
Synthetic varnish	180	
Silicone varnish	200	

^{a} Maximum continuous operating temperature.
^{b} The ultimate temperature at which the insulant can function under overload conditions without complete failure.

TABLE 10

CABLE INSULATION: RESISTANCE PROPERTIES IN HOSTILE ENVIRONMENTS

Oil	PTFE, PVC, polypropylene, polyurethane, nylon
Weather, sun	CSP, PTFE, silicone rubber, polypropylene, nylon
Abrasion	Rubber, nylon, ETFE, polyurethane, Kapton
Nuclear radiation	Silicone rubber
Water	PVC, polypropylene, nylon, ETFE
Acid	CSP, polypropylene, PTFE
Alkali	CSP, polypropylene, nylon, PTFE
Aliphatic hydrocarbons (e.g. petrol)	Nitrile rubber, PTFE
Aromatic hydrocarbons (e.g. toluol)	PTFE
Halogenated hydrocarbons (e.g. degreaser solvents)	PTFE

number of insulation materials which exhibit good resistance to the indicated hostile environments.

Although PVC-covered cable is probably the most readily available and is cheapest and consequently most widely used, it should be used with caution outside 'ideal laboratory' conditions. The quality can vary considerably between manufacturers, the lower quality PVC often being porous and the quality of insulation in different batches of the same type of cable from a single manufacturer can also vary. However, the higher quality PVC-covered cables can be used in certain hostile environments, for example, underwater (static, flowing and under pressure). In all cases where cable is intended for use underwater, it is strongly recommended that it is pressure tested or immersed in water and the resistance-to-ground (RTG) value measured to check for pinholes in the insulation. Failure rates up to 10 % can be expected. One distinct advantage of PVC-covered cable is that it is readily available in a wide range of types, e.g. single core, multicore and ribbon cable, and in many colour codes. The effects of environment on PVC cable insulation are illustrated in Figs 3 and 4. Figure 3 shows the effect of heat on the conductivity of PVC. As the temperature rises, the PVC insulation value drops, such that at 100 °C there is a 10 megohm path between the conductor and the outer surface of the cable—the value rises again to 10^4 megohms as the temperature is reduced. This effect is continually repeatable for any change in temperature. Figure 4 shows the effects of applying paint containing MEK to PVC-covered cable. The first application results in a rapid drop in the insulation value and the second and third applications also have a detectable effect. (The PVC is permanently affected by the MEK, causing it to swell.) The insulation resistance value of the covering does not return to its original value until four months after paint application, and permanent damage is caused.

Solder

Selection of the correct solder for a particular application is based almost entirely upon the operating temperature and time-at-temperature of the gauge installation. It must be emphasised that only 'strain gauge solders' (i.e. those recommended by strain gauge suppliers) should be used in every situation—many commercial solders contain highly active fluxes which are not suitable for strain gauge applications. A separate corrosive flux must *never* be used.

Three basic types of strain gauge solder are available. Firstly, a general type, with suitable characteristics of high electrical conductivity, good

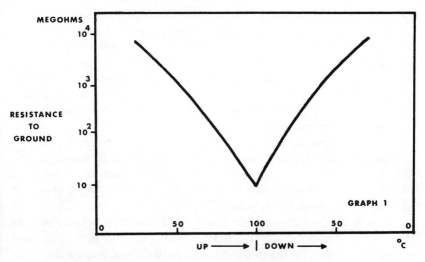

FIG. 3.   Effect of heat on PVC-covered cable.

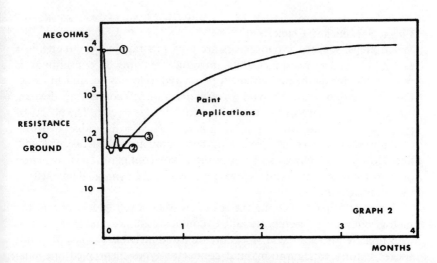

FIG. 4.   Effect of MEK on PVC-covered cable.

'wetting' and flow, corrosion resistance and mechanical strength. This type is based on a 63 % tin/37 % lead alloy, giving a melting point of approx. +180 °C, but should not be used below −7 °C for long periods as phase transformation of the tin (termed 'tin disease') may occur, leading to failure of the joint. To overcome this problem a second type, containing a small percentage of antimony (≃0·35 %) is available, permitting measurements at cryogenic temperatures. A third group of alloys is available for high temperature applications with typical melting points in the range +220 °C to +310 °C. More detailed information is given in reference 5.

Conventional soft solders are normally limited to a maximum operating temperature some 30 °C below the melting point, to ensure that sufficient strength is retained in the joint: the highest practical operating temperature is therefore in the order of +280 °C.

Between approximately +260 °C and +620 °C there is a choice between silver soldering and spot-welding. Silver soldering is usually carried out with a resistance soldering tool and requires special lead preparation.[6] The solder used has a melting point of approximately +620 °C and is often supplied in the form of a powder ready mixed with a suitable flux. Above +650 °C there is no alternative to spot-welding which is normally carried out with a capacitive discharge welder equipped with a hand probe and variable energy control. This technique requires the provision of specialised lead materials, cables etc.

Plugs, Sockets and Connectors

When plugs, sockets or connectors are used in strain gauge circuits, it is essential that the resistance they introduce remains as constant as is practicable during the measurement phase and as low as possible in value. Great attention must be paid to the selection of connecting devices, especially if they are within the Wheatstone bridge circuit. The effects of plugs, sockets and connectors is greatest when a $\frac{1}{4}$ bridge two-wire configuration is employed, because of the accumulative effect of each signal line. However, the effect in a $\frac{1}{4}$ bridge three-wire configuration is mismatch of the resistance values (and changes in them) for the two signal lines within the bridge circuit.

The main criteria affecting the selection of connecting devices and the factors which affect the repeatability of contact resistance (and changes in it) with time, temperature and vibration are listed in Table 11. The plug and socket mating configurations and connector types mentioned are illustrated in Fig. 5.

TABLE 11

STRAIN GAUGE CONNECTOR SELECTION CRITERIA

	Practical considerations	S/G applications
Plugs and sockets		
Round pin	Precision machining necessary to obtain maximum use of contact material and repeatable, stable contact resistance.	Rarely cost effective
Flat pin	A wide range of designs available. A bifurcated construction is the superior type.	Bifurcated type
Leaf spring	Provides a quick release system, but low use of contact material. Contact resistance varies with age—unsuitable in high vibration situations.	Rarely suitable
Helical spring	Provides a good connection once made. Unsuitable for repeated make-and-break situations.	Limited
Crimp	Two types available, single and four-point; special tool required for both. Four-point generally superior, as tool has controllable crimp action, ensuring all round compression of cable.	Four-point recommended
Connectors		
Taper-pin	Requires special tools to make-up, insert and withdraw but provides a good connection once made. Unsuitable for repeated make-and-break situations.	Limited
Barrier strips	A wide range of designs available, varying from poor to sound mechanical construction. A screw barrier strip provides a good connection system, with visual checking facilities.	Screw barrier strip satisfactory
Clip-on	On no account to be used within a S/G bridge circuit as they provide non-repeatable, varying contact resistance in almost every situation.	NOT recommended

Fig. 5. Strain gauge plugs and sockets and connectors.

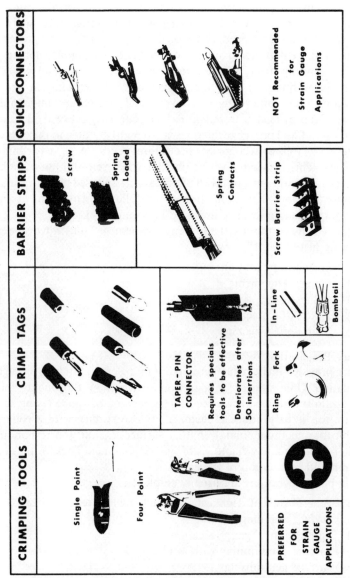

FIG. 5.—contd.

INSTALLATION TECHNIQUES

A hostile environment during the gauge installation process can have a marked effect upon the usable life of a gauge installation and directly affect the validity of the data produced. Consideration must also be given to the effects of the environment on the operator. For example, human limitations resulting from stress caused by extremes of ambient noise, temperature, humidity and dirt, together with working in cramped conditions, may often render conventional techniques impossible to achieve. Failure of gauge installations using correctly-selected high-quality materials and well proven techniques is often attributable to the operator's inability to follow recognised procedures because of environmental restraints. This illustrates yet again that when selecting an installation technique, account must be taken of limitations imposed by the installation environment and that simulation of installation environment in the laboratory must be realistic if it is to be meaningful.

Some effects of the installation environment are illustrated under the following headings:

(1)   Surface Preparation;
(2)   Gauge Alignment;
(3)   Adhesive Application and Curing;
(4)   Lead Attachment;
(5)   Technique Checking.

Surface Preparation

Surface preparation is a vital factor in obtaining credible data from a gauge installation. Unless surface preparation is carried out thoroughly, incomplete strain transmission can occur. The purpose of gauge laying surface preparation is to produce a chemically clean surface with a roughness suitable for maximum bond strength, a surface alkalinity corresponding to a pH of approximately 7, and visible gauge alignment lines for the correct location and orientation of the gauge.[7]

Five basic operations are usually employed:

(1)   general solvent degreasing of the gauging area;
(2)   surface abrading;
(3)   application of gauge location layout lines;
(4)   surface conditioning with a mild acid;
(5)   surface neutralising.

The initial condition of the surface of most materials to which gauges must be applied on site is often rusty or painted and is invariably pitted, necessitating some form of mechanical abrasion. Access can be a major problem and it is not uncommon to find that it is only possible either to insert a grinder *or* to look at the gauge area, not both at once. The best advice that can be given is to do the best that environmental restraints permit and to make a detailed note of the condition of each area, as this may subsequently provide valuable information as to the reliability or otherwise of the data obtained. Where possible, an area large enough for the gauge and protective coating should be mechanically cleaned.

A minimum requirement is to obtain a bright metal surface, and where the surface is pitted a wire brush will often help. The final condition achievable by mechanical abrasion will be a deciding factor in the use of a particular strain gauge adhesive. Also, it may be found that the original intention to employ a 'thin glue-line' adhesive system is just not practicable, owing to the pits remaining under the gauge grid. If it is not possible to reposition the gauge to a pit-free area, it will be necessary to use 'thick glue-line' techniques and may in addition necessitate a re-examination of the gauge excitation levels to ensure that self-heating of the grid does not occur because of the inferior heat transfer qualities from the grid through the thicker adhesive layer.

In conditions of restricted access, it is obvious that the operation requiring the most dexterity, i.e. the application of the gauge alignment lines, is the most difficult to achieve; indeed, it is often physically impossible to make a satisfactory job of this essential operation. Again, the only advice is to achieve the best that is humanly possible, and to note all the relevant factors. One technique that has been successfully employed is to align the gauge as well as possible and then take a photograph, showing some identifiable part of the structure that will enable a measurement of the amount of misalignment to be determined, thus enabling corrections to be made to the strain readings.

Restricted access can also affect the thoroughness of chemical surface preparation, where operator frustration, resulting from cramped conditions, unnatural physical position and sheer exhaustion tend to result in departures from rigid procedures. It is very much a case of 'mind over matter'. Particular attention should be paid to avoiding excessive use of metal conditioner, contamination of cleaning materials and to ensuring complete neutralisation of the surface.

Gauge Alignment
Gauge grid orientation is critical as this has a direct effect upon the level of surface strain to which the gauge will respond and therefore affects the

related computed stress. For example, in a plane uniaxial stress field, at approximately $62°$ (depending on the value of v) the measured strain would be zero, yet 25 % of the maximum normal stress is present. It is, therefore, essential to recognise that for a uniaxial stress field, the relationship $\sigma = E\varepsilon$ is only valid along the direction of the applied load and the gauge axis should be aligned with this. Any misalignment will result in a reduction in the strain response of the gauge and will necessitate the use of a correction factor when computing the corresponding stress (see Chapter 5). A similar problem also occurs in a biaxial stress field. Reference 8 provides more detailed information.

As already stated, physical limitations may severely affect the accuracy with which the gauge alignment marks can be applied, or may even prevent their application. Where it has been possible to mark the surface, it is essential to have good lighting conditions to enable the gauge to be aligned correctly to the marks. In the majority of cases it will be found essential to apply the gauge using two hands, with the gauge secured to a piece of suitable transparent tape and where a terminal is required it is advisable to bond this at the same time as the gauge. Again, limited access may render this or subsequent operations impossible. One technique that has been successfully employed to overcome access problems is to pre-encapsulate the gauge (as described on page 128), which can then be installed single handedly. As stated, a photograph of the installed gauge will enable any necessary corrections to be made.

Adhesive Application and Curing
It is strongly recommended that a trial run of all the operations, with all the actual equipment, is carried out prior to the gauging exercise to ensure that the installation environment has no detrimental effect on the gauge bonding process. Where possible, this should include measurement of glue-line temperature and verification of clamping pressure, where these are applicable to the adhesive being used. Each type of strain gauge adhesive has a particular cure cycle, based on time, temperature and pressure factors. These cure cycles, and the shelf and pot lives of the adhesive have been derived by the gauge manufacturer to ensure that the best possible qualities are obtained. This information was obtained in 'ideal laboratory' conditions and is, to a certain extent, environmentally-dependent. The recommendations should, however, be strictly observed—within the restraints of the installation environment.

The use of incorrect or non-uniform clamping pressures can induce large residual stresses into the cured adhesive layer, which will vary with time, strain level and temperature, making the gauge installation unreliable and the data obtained not repeatable (see Chapter 5). If the stresses are large, superimposed strain transfer stresses drive the adhesive into a markedly non-linear portion of the stress/strain curve, resulting in a lowering of the incremental gauge factor and increasing non-linearity of the grid output, even though the grid itself is still in the linear strain resistance portion of its own characteristic. The situation is further complicated by the effects of surface preparation on individual installations and the possibility of residual strains being induced into the adhesive during the curing cycle.

Lead Attachment
As described earlier, there are three basic methods of lead attachment and the major selection criterion is the measurement temperature, though other environmental restraints may apply. For example, in nuclear radiation fields, only spot-welding is recommended.[9] The most commonly employed technique is soft soldering, which will be considered in detail.

Lead attachment is invariably another two handed operation and access to the gauge area can be a major limitation. Physical comfort, access at the right working angle, high ambient lighting and the right mental approach are vital factors in the success or otherwise of making satisfactory lead attachment, as is the availability and correct use of the right tools: high quality cutters, wire strippers, tweezers, scissors, pliers, dental probes, scalpels etc. and, probably of greatest importance, the soldering equipment essential to make a reliable lead attachment that will withstand the hostile environment.

An important consideration in lead attachment is the lead configuration employed. In a large number of hostile environment applications multi-grid gauge patterns (e.g. three grids at 45°, rosette) are necessary to determine the magnitudes and directions of the principal stresses, and it is essential, when using a Wheatstone bridge configuration energised by a constant voltage (CV) supply, that a $\frac{1}{4}$ bridge three-wire configuration is employed. The four different methods of lead attachment for a $\frac{1}{4}$ bridge connection using a terminal are shown in Fig. 6 and it will be noted that the only configuration which is entirely free from the effects of temperature on the jumper leads and signal cable is that which forms a 'true three-wire' system right up to the gauge solder pads. This method is therefore recommended where there is access for it and in those situations where access is limited the right-angled terminal illustrated can prove more suitable. Use of the gauge configuration option where the jumper lead is an integral part of the gauge, can cause problems in temperature applications, as a small part of the measurement system is in 'two-wire' format.

Another major consideration is the signal cable routeing between the gauge and the rest of the bridge completion network, where the signal cable is inevitably part of the bridge circuit. The cable should be routed in such a way that the effects of high electromagnetic and electrostatic fields and local areas of high temperature are kept to a minimum, and that with a $\frac{1}{4}$ bridge three-wire system, these effects are common to the two leads in the bridge circuit. The cable should also be satisfactorily anchored, to reduce the effects of cable movement which causes triboelectric noise and insulation chaffing. There is a tendency, where suitable three-core cable is not available to use a two-core cable together with a single-core cable.

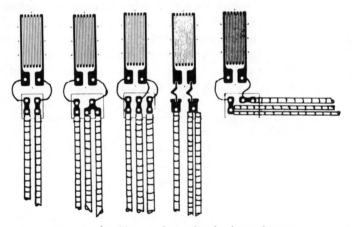

FIG. 6. $\frac{1}{4}$ bridge configuration lead attachment.

When this is the case, it is essential that the two-core cable is connected to the correct arms of the bridge (see Fig. 7A) to ensure symmetrical cancelling properties in the three-wire configuration. It is all too easy (particularly as it looks neater) to make the connections as in Fig. 7B, without realising the full implications.

Considering that the actual soldered joint is within the bridge circuit, it is a vital part of the strain measurement system; any failure will result in complete loss of data, and any variation in resistance will result in a false strain reading, indistinguishable from that due to surface strain. Great care should be exercised in preparing for and making the soldered joint, the major pitfalls to guard against include:

(1)   incomplete removal of enamel-type insulations;
(2)   damage to or severing of cable conductors when removing outer insulation;
(3)   sharp bends or nicks in jumper leads;
(4)   soldering-iron temperature incorrect;
(5)   flux burnt out of solder before application to joint;
(6)   lead movement during cooling phase of soldering;
(7)   'dry' soldering-iron bit, causing solder not to flow properly.

The importance of the use of correct strain gauge solder has already been considered. However, it is important to realise that the melting point of the solder is governed by the lead/tin ratio and any increase in the tin content reduces the melting point. Failure of a high temperature strain gauge installation due to melting of the soldered joint could be caused by an increase in the small tin content ($\simeq 5\%$) of a high temperature solder by a small amount of 63% tin/37% lead solder remaining on the soldering-iron bit or being picked up from a soldering-iron sponge, thereby increasing the tin content.

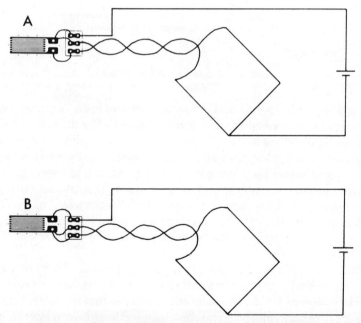

FIG. 7.   Three-wire bridge configuration lead attachment.

Turning our attention to the mechanics of producing a soldered joint that will prove reliable in a hostile environment, consideration must be given to the soldering-iron used. The temperature of the soldering-iron bit should be such that there is satisfactory melting of the solder in the actual environment in which it is to be used, taking into account the ambient air temperature, effects of draughts and the heat sink effects of the material to which the gauge is bonded. A temperature controlled iron with a pretinned iron-plated tip is probably the most suitable type. When the signal cable is stranded, it is often convenient to use one of the strands as the jumper lead, cutting off the remaining strands approximately 2 mm from the insulation, twisting the ends together tightly and then soldering to the terminal at this point (see Fig. 8).

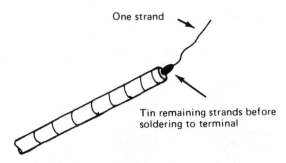

One strand

Tin remaining strands before soldering to terminal

FIG. 8.   Single-strand jumper lead.

It is essential for production of a satisfactory soldered joint that there is no movement of the items to be soldered together while the solder changes from its liquid phase to its solid phase. One recommended technique to ensure this is to tape the signal and/or jumper leads into their final positions prior to soldering. This practice leaves both hands free to apply the solder and the iron together, which is essential to reduce the possibilities of (a) the flux in the solder being burnt out before it reaches the joint, (b) a spike of solder being formed as the iron is removed from the joint. The former can result in a 'dry joint' and the latter can reduce the effectiveness of the protective coating.

There are two schools of thought regarding the presoldering of jumper lead, gauge solder pad and terminal prior to final soldering. Presoldering can result in excessive use of solder and may mean that the solder becomes part of the measurement circuit rather than serving solely to keep the lead wire in contact with the terminal. Care should be taken to avoid this.

However, taking all things into consideration, the author prefers the presoldering technique, as it ensures that all items are adequately wetted with solder, reducing the possibility of a 'dry joint', particularly in the area that cannot be inspected, i.e. where the two surfaces meet.

Certain types of gauges have pretinned gauge tabs (often termed 'solder dots') and different lead attachment techniques are required for these.[10]

In a number of cases, in particular when the adhesive has been heat cured, it will be found that the copper terminal tab has oxidised, preventing adequate wetting by the solder. This oxide layer can be successfully removed by lightly rubbing the terminal with a hard pencil eraser. To ensure maximum solder wetting of the terminal pad, it is recommended that the terminal is cleaned in this way in every case.

After completing the joint, it is necessary to remove the flux residue to ensure a high 'resistance-to-ground' (RTG) insulation value and to eliminate the possibility of long-term chemical attack of the soldered joint and/or gauge grid by any remaining active flux. A number of proprietary brands of 'rosin solvent' which have no detrimental effect on strain gauge installations are available. It is recommended that a stiff brush is used, with sufficient solvent to dislodge hard residue and that dislodged residue is washed from the gauge area. It must be continually borne in mind, however, that the solvent may be inflammable or toxic and care should be exercised, particularly in confined spaces. For safety reasons, it should, of course, only be used from approved containers. To verify that all the residue has been removed and that all the solvent has evaporated, it is necessary to measure the RTG value with a suitable low voltage insulation tester. In certain situations, up to 30 mins may be necessary for all the solvent to evaporate from inside the cable insulation. Before applying any protective coating, it is advisable to check that RTG values of greater than 10^4 megohms can be achieved.

Special precautions for lead attachment are necessary when the strain gauge is to experience high centrifugal forces,[11] and when the maximum fatigue life of the strain gauge is to be realised.[12]

Technique Checking

One recommended method of verifying that the correct materials and gauging techniques are being employed in a specific hostile installation environment and that the installation should provide viable strain measurement data, is to lay a gauge on a gauge factor test bar (see Chapter 5, Fig. 15). Preferably this gauge should be of the same batch as that being used in the measurement environment, but certainly it should be laid by the same

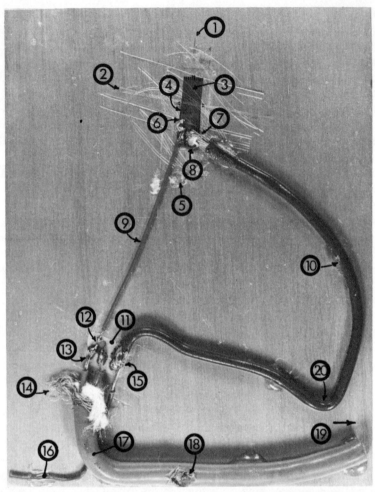

FIG. 9. Twenty principal installation and cable connection mistakes. 1: Indistinguishable axial alignment line; 2: transverse alignment line not at right angles to the axial one; 3: area under and around gauge still contains score marks; 4: too much catalyst; 5: too much adhesive; 6: solder shorting-out grids; 7: strands of jumper lead not soldered onto terminal; 8: solder not flowed properly; 9: jumper lead too tight, no stress relief loop; 10: dissimilar length and size of jumper leads; 11: terminal too far from gauge; 12: insufficient use of terminal area; 13: potential track path; 14: unprotected screen causing intermittent earthing; 15: 'dry joint' producing tensile strain indication; 16: single lead in incorrect leg for $\frac{1}{4}$ bridge three-wire; 17: signal leads crossed; 18: outer insulation damaged; 19: signal leads in three-wire configuration following different routes; 20: stress relief loop bend too tight. Courtesy of Vickers, Shipbuilding and Engineering Limited.

operator, using the same materials, techniques and equipment, and in such a way that the full effects of any restraints of the installation environment are reflected. There should, therefore, be no simulation or 'cheating'; everything should be done as though it were the actual job, at the actual job location, and preferably in such a way as to reflect the worst conditions from the gauging point of view. The gauge factor of the laid gauge should be checked as described in Chapter 5. Additional information can also be obtained in those situations where it is practicable to subject the bar to the measurement environment (actual or realistically simulated) by noting the effects on the soldered joint, adhesive etc.

To conclude this section on installation techniques, due consideration should be given to Fig. 9, which illustrates twenty of the principal mistakes that can be made when installing and connecting the gauge to the signal cables. Some of these mistakes are all the easier to make in a hostile environment and must be all the more zealously guarded against. Certainly the effects of a number of them will be greater in a hostile measurement environment.

PROTECTION

It is probably true to say that of all the aspects of strain measurement in a hostile environment, protection of the installation from the environment, both during installation and measurement, receives the least consideration and, consequently, is a major cause of failure. Two points are evident: (1) the philosophy of gauge protection and protection techniques, unlike, say, gauge laying techniques, is not well documented or even adequately discussed; (2) satisfactory gauge protection can only be achieved by careful thought and preparation *prior to and during* the gauge laying process, and *not* afterwards, as is often the case. The author's philosophy is expounded in detail in reference 13.

Before considering in detail the practicalities of protection in a hostile environment, it is essential to appreciate some of the basic principles and something of the associated philosophy. The first consideration is 'Why is it necessary to protect a foil gauge from its surrounding environment?'. Three basic reasons are evident: (1) to protect the adhesive, and thus retain its strain transmission capability; (2) to provide physical protection to the delicate gauge grid and jumper leads; (3) to prevent apparent strain readings due to (a) chemical attack of the gauge foil and soldered joint etc. which would result in false tensile strain readings, and (b) insulation leakage paths, producing 'shunting' of the gauge resistance either directly or between the gauge and the metal structure to which it is bonded, either of which will result in false compressive strain readings. For example, assuming an initial balance value of 10^4 megohms, 5 microstrain (compressive) will be indicated if the d.c. RTG value falls to:

12 megohms (120 R gauge)

35 megohms (350 R gauge)

Considering a typical gauge installation, Fig. 10 shows the three major breakdown paths for the ingress of fluids in wet environmental conditions and which may result in chemical attack or insulation breakdown. The first path is at the structure/coating interface and is directly related to the quality of the bond of the protection material to the structure. The 'goodness' of the bond of the protective coating to a metal structure is dependent to a large extent on the amount of time and effort that is given to surface preparation. For long-term installations particular care and attention in surface preparation is as necessary for gauge protection as it is for gauge bonding. Ideally, surface preparation, both mechanical and chemical, should be performed for both gauge laying and coating application at the same time, with the shortest possible delay between the two operations. In practice this ideal is difficult to achieve. There is little documented evidence to support the arguments as to whether a rough or a smooth surface finish produces a better bond. Experience has shown that it is very much a question of the nature of the materials concerned. For example, it can be shown that smooth wooden surfaces produce stronger joints than roughened ones and that the bond strength of rubber to textiles depends upon the number of fibres which are embedded in the rubber, and is unaffected by surface treatments.[14]

Adhesion is a complex process, probably involving mechanical interlocking, molecular absorption, interdiffusion and electrostatic attraction factors, which vary for each particular substrate and adhesive combination. It is evident, therefore, that the best surface preparation for a

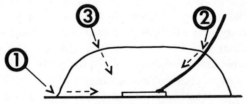

FIG. 10. Gauge installation breakdown paths.

particular combination, particularly surface roughness, should be determined experimentally. In the author's experience, the most critical factor in surface preparation for strain gauge protection applications, is the selection and sequence of application of chemicals. Where surface roughness has been demonstrated as being non-critical, a smooth surface is preferred, as it permits easier chemical surface preparation. It can be demonstrated that immediately after chemical cleaning with metal conditioner and neutraliser, a 'water drop test' exhibits the expected characteristics of 'an even flow from the surface without forming droplets or discontinuities'.[14] If this test can truly be considered a valid indicator of optimum bonding conditions, then it appears that prior to the application of a protective coating material, chemical cleaning should be carried out. An interesting observation is that the water drop test shows an apparent deterioration of the surface conditions with time (e.g. after 15 min in laboratory conditions) as indicated by changes in the flow pattern. Above all, surface preparation must be such that it enables good wetting of the structure by the coating material, as this is a prerequisite for good adhesion. An additional point, which is thankfully easier to understand, is that it is necessary to degrease all cable outer insulation which comes into contact with the protective coating, in order to ensure a better bond.

The second breakdown path is the cable/coating interface. When selecting cable type, consideration must be given to the bonding characteristic of the proposed coating material to the cable outer insulation. It can be demonstrated experimentally that even the pigments used to produce PVC-insulated cables in different colours can have a marked effect upon the effectiveness of the bond between certain coating materials and the PVC. For example, failure of the coating bond to white coloured PVC insulation will occur sooner than failure of that to black. PVC-covered cable must be used with caution on long-term installations (i.e. over one year) in wet conditions as it is heavily plasticised to produce flexibility, and in air or fluid flow conditions the plasticiser tends to separate from the PVC, which then reverts to its natural rigid form, eventually cracking and permitting ingress of moisture etc.

The third breakdown path is through the coating material itself and the danger of this can be partially reduced by selecting a material which exhibits low moisture transmission and absorption properties. The major problem is in the application techniques which must ensure that there are no air-pockets or tracking paths present in the coating.

A very important aspect of protection is the coating material itself. Indeed, it is often true that 'the material makes the technique possible'. In many cases it will be found necessary either to evaluate the effects of a specific hostile environment upon a recognised strain gauge coating material, for instance where the manufacturer does not specify its use in the particular situation, or, as is often the case, to evaluate a material that is recognised for use in the particular environment but is not generally recognised as a strain gauge coating material, and therefore its effects on a strain gauge installation are unknown. A good example of the latter case are certain room temperature curing (vulcanising) RTV silicone rubbers, which have good high temperature properties, but liberate acetic acid during the curing process—this can completely destroy a gauge foil.

Protective coating materials should have the following properties: they should

(a)   be free from moisture, active acids and salts which can cause grid etching;

(b)   be free from solvents which are incompatible with the gauge adhesive used;

(c)   be pliable over the required temperature range, and cause no local stiffening of the structure;

(d)   adhere tenaciously to both metal and cable;

(e)   provide mechanical protection;

(f)   have low water absorption;

(g)   be easy to handle and apply, i.e. no primers;

(h)   have correct viscosity, e.g. non-slump;

(i)   not produce toxic fumes if used in confined areas;

(j)   be safe and practicable to handle in the installation environment;

(k)   be 'cold curing' if used on large structures in the open;

(l)   be 'catalyst curing', if used for high build-up applications and exhibit suitable characteristics for installation time and conditions (e.g. temperature and humidity).

Where it is necessary to evaluate possible coating materials the adoption of a selection test programme is essential. The environment in which the gauge is to be installed and operate should dictate the programme format and the testing methods adopted. Some selection and evaluation programmes have been reported.[15,16]

The one described here is a three-stage selection test programme (see Table 12) which was developed by the author and has been used for a number of years to evaluate possible new strain gauge coating materials; to 'quality control' materials in general use; and to determine the long-term

TABLE 12

THREE-STAGE PROTECTION SELECTION TEST PROGRAMME

Stage one	Initial selection: Laboratory tests—checked weekly	Small samples used for (a) initial product selection (b) quality control on products in general use (batch tests) (c) additional product evaluation (mixing ratio and surface preparation tests)
Stage two	Detailed testing: Laboratory simulated environment tests—checked monthly	(a) further bond tests (larger samples), (b) water absorption, (c) cable bonding, (d) grid etching (zero drift), (e) simulated environments: (i) driving rain; (ii) cyclic strain; (iii) temperature; (iv) pressure; (v) flow; (f) effects on gauge strain response
Stage three	Field testing: 'Real-time' tests—checked monthly	Conditions to suit usual test environment, e.g. shipbuilding (a) pressure and flow (b) outside weathering (samples at 45° facing south) (c) cyclic tests with full/partial immersion, with sun, rain, wind, frost and tidal flow (d) total immersion, sea water

effectiveness of particular protection techniques.[17] Although this programme was specifically devised for a shipyard situation where heavy engineering processes and partial and full immersion in freshwater and seawater environments predominate, the general principles and philosophy apply to most situations.

The basic requirements of an effective test programme are that it must be:

(a) simple—able to be performed by laboratory assistants;
(b) cheap;
(c) realistic—in relation to the installation and measurement environments;
(d) organised—well documented;
(e) versatile—to pursue test findings and accommodate changes.

A successful programme should reduce the laboratory evaluation time in relation to the actual field 'real' time. For example, a suitable time compression would be one month laboratory evaluation equivalent to one year field life.

Crucial to the success of the test programme is the method of detection of material failure and the relation of this to the installation and measurement environments. For example, a large proportion of long-term marine strain gauge installations fail because of ingress of water, and any selection method must detect this breakdown. In the opinion of the author, and in accordance with the requirements of the test programme, a change in d.c. resistance-to-ground (RTG) value is the simplest and most cost-effective method of determining this breakdown, provided that the on-site measurements are to be made with a d.c.-excited bridge instrument. For an a.c.-excited measurement system, RTG measurements are not sufficient[18] and a capacitive reactance measurement method, at the bridge supply frequency, is necessary.

The initial stage of the three-stage selection programme is used to determine if a material has the potential for use as a coating material for strain gauge protection. Small samples of the coating material are bonded to steel slips (approximately 30 mm × 13 mm), precleaned in accordance with normal gauge laying practice. Strain gauge terminals are used instead of gauges, which considerably reduces costs. A 7·020 PVC-covered lead is soldered to the terminal, the joint is defluxed and when the RTG value is greater than 10^4 megohms, the prospective coating material is applied. When the coating is fully cured (usually, though not always, indicated by the

RTG value returning to greater than 10^4 megohms) the sample slip is fully immersed in the relevant test fluid (e.g. tap, rain or seawater) and RTG readings are taken and recorded weekly: a reduction in RTG value indicates ingress of fluid and therefore the onset of failure.

The second stage of the test programme performs a number of functions. It provides a more detailed evaluation of those materials which, at the initial stage, have exhibited the required properties for a coating material, i.e. in the case of marine applications, the ability to resist moisture ingress. Larger test samples, of a more realistic size in relation to the actual installation, are applied to test plates of the structure material and strain gauges are used at this stage to enable the effects of the coating material on the grid foil and the cured adhesive to be studied by measuring gauge resistance. (This is a good use for the odd assortment of gauges that are inevitably left from previous measurement exercises.) It is essential that the test samples are truly representative of the exercise installation. For example, if three-grid rosette gauges are to be employed in a three-wire bridge configuration then nine separate cables should emerge from the coating, as breakdown at the cable/coating interface is the most common cause of installation failure.

Detection of potential cable/coating interface failure is another important function of stage two, and is achieved by cable bonding tests. The ends of precleaned cables are protected with the coating material and, after curing, are immersed in the appropriate fluid. Failure of the cable/coating interface is again detected by a reduction in the initial RTG value. This technique is also employed to evaluate the effectiveness of the coating bond to different types of cable insulation, and the effect on a cable/coating bond of any production changes by the cable manufacturer.

The second stage of testing can also be used to demonstrate the effects of different surface preparation techniques and the effects of different mixing ratios on bond effectiveness. In addition, different test environments can be simulated, to check the effectiveness of the proposed coating material/technique combination. In most cases, with a little ingenuity and readily available facilities, a cheap yet effective simulation can be achieved. For example, damp conditions (the effect of which is often more detrimental than full immersion) can be simulated with a coffee jar, some clean sand and a length of plastic conduit. The conduit is placed in the middle of the empty jar, the jar is filled to about one-third with sand, the test sample is inserted and then the jar is completely filled. The plastic conduit is then filled with the test fluid and, if topped up periodically, damp conditions are maintained. Similarly, driving rain can readily be simulated by blowing air (e.g. works air supply) across a tube immersed in water, the complete system being enclosed in a water tight box, preferably with a transparent side for observation.

The third stage of evaluation consists of 'real-time' testing and ideally should be concluded prior to purchasing the materials for the measurement exercise. In many cases, however, time scales do not permit this ideal to be realised and it is not unusual for the decisions on materials and technique to be made during stage two (another good reason why stage two should be realistic). However, stage three is a very important part of the evaluation programme, as it provides data on the survival of the protection technique in the actual environment. Although it will obviously take five years from the commencement of the test to actually verify that a certain material/ technique combination will survive a particular hostile environment for five years, in the author's opinion, there is no substitute for 'real-time' testing, even though it can often prove expensive if test facilities need to be manufactured.

Use of the test programme outlined has resulted in a number of non-recognised materials being found suitable for strain gauge protection in various hostile environments, and a list of some of these, with additional information and typical applications, is given in Table 13.

When considering practical techniques for the protection of a gauge installation, it will probably be found that there is no fully documented technique 'tailor made' for the particular hostile environment under consideration. Each individual installation imposes its own limitations and may necessitate changes in materials and techniques or even the development of new methods.

There are certain basic criteria for the selection of protection techniques:

(a) technique practicable in time and environment;
(b) consideration of extremes of environment during installation, premeasurement and measurement phases, including time factors;
(c) local stiffening of the structure caused by protection technique;
(d) cyclic effects of strain, temperature or pressure during measurement;
(e) amount of surface preparation practicable (mechanical and/or chemical);
(f) accessibility affecting ease of application;
(g) quality inspection during application;
(h) availability of high quality, batch tested materials;
(i) proven effectiveness of technique selected;
(j) cost-effectiveness.

TABLE 13
STRAIN GAUGE PROTECTION MATERIALS

Product	Type	Quoted °C temperature range	Shelf life at room temperature (months)	Mixing ratios	Cure times at room temperature	Typical application
Berger chemicals PR 905	2-part polysulphide epoxy	−55 to +150	6	A 12 B 11 By weight or volume	Pot life ≈30 min Full strength 24 h	Used to pre-encapsulate gauges. Gives long-term protection in sea water.
Bostik 2114-5	2-part polysulphide synthetic rubber	−50 to +130	6	Part 1 10 Part-2 1 By weight	Pot life 5–8 h Full strength 7 days	All round general purpose. Ideal for most water immersion conditions. Provides good mechanical protection.
Bostik 771	1-part nitrile rubber/resin	−40 to +100	12		Full strength 24 h	Provides effective method of PVC cable repair and protection of terminals and tags. Brushable.
Ciba-Geigy AV 138	2-part epoxy resin	−55 to +120	12	AV 138 HV 998 10 4 By weight	Pot life ≈20 min Full strength 48 h Full strength 72 h	Gives good mechanical protection and is effective in hot oils. Useful for anchoring lead wires and cables.
Dow Corning RTV 738	1-part silicone rubber	−60 to +180 (up to +260 intermittently)	6			Suitable in adverse construction conditions, i.e. preheating and welding, grit blasting, etc.
Fleming services GC 101	2-part polysulphide cured epoxy	up to +75	12	GCR 101 GCH 101 1 1 By weight or volume	Pot life ≈30 min Full strength 24 h	All round general purpose, especially where tough mechanical protection required.

The principle of checking the effect of the installation environment that is established for the gauge laying process (Chapter 5) is equally relevant to the protection technique. The proposed protection technique should therefore be evaluated by preparing a test sample in typical installation environmental conditions and then subjecting it to a realistic simulation of the expected measurement environmental conditions (i.e. stage two of the test programme) ensuring, in particular, that post gauge installation/pre-measurement environmental conditions are experienced by the test sample. One common mistake is to evaluate a protection technique as described above and then, prior to the measurement exercise, change the technique or the materials, without realising the full implications on the effectiveness of the protective coating, and, therefore, not retesting the modified technique. The converse is also true of course, that if the hostile environmental conditions are changed, e.g. owing to a change in a manufacturing process, then it may be necessary to recheck the effectiveness of the protective technique/materials.

Whatever technique is employed, there are a number of basic protection requirements which, if given careful attention, will ensure maximum effectiveness of the protection. Table 14 lists these basic requirements, as a series of procedures.

A number of techniques developed by the author for strain gauge protection in a shipyard environment are described. These techniques

TABLE 14
BASIC STRAIN GAUGE PROTECTION REQUIREMENTS

1	Mark the area around the gauge to be covered, before obscuring the gauge.
2	Measure and record gauge grid resistance and resistance-to-ground (RTG) after removal of soldering flux but before applying coating.
3	Mix two-part materials thoroughly and beware of unmixed material on the container surfaces.
4	'Wet' the surface with a small amount of material before commencing full protection.
5	Do not feather the protective covering but have a high build-up at the edge.
6	Avoid sharp corners at the coating edge, as breakdown can begin at such places.
7	Remove, where practicable, possible tracking breakdown points from the structure surface (e.g. terminals).
8	Use pressure-tested cables and anchor them, both inside and outside the coating area, before coating.
9	Degrease all cable cores in contact with the coating and ensure separate entry into the coating for each one (this applies also to ribbon cable).
10	Make the cable/coating interface as long as possible.

provide mechanical protection and prevent moisture ingress to the gauge, in either full or partial immersion conditions.

Short-term Technique (up to 6 months' protection)
When a relatively short-term protection technique is required, up to 6 months in wet conditions, that illustrated in Fig. 11 has proved to be satisfactory. Provided that the basic protection requirements (Table 14) are observed, this technique will also provide adequate gauge protection at pressures up to 2500 lbf/in^2. Note the extended length of the cable/coating

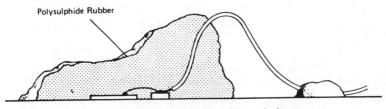

Fig. 11.   Short-term protection technique.

interface and the method of cable anchorage outside the coating area to reduce the possibility of the cable/coating bond breaking at the interface. When the technique is employed in pressure applications it is essential that the coating is applied void-free and that pressure tested cable is used. To increase the effective life of the protection, a paint, compatible with polysulphide rubber and the cable insulation, can be used to prevent oxidation at the structure/coating interface.

Over-lap Technique (up to 12 months' protection)
This technique differs fundamentally from the 'short-term' technique in that it is a 'double barrier' technique, although both barriers can be of the same material (see Fig. 12). The cable/coating interface is double that of the 'short-term' technique, which increases the life of the installation. (A number of coatings fail at this point.) Using the 'over-lap' technique, adequate gauge protection in wet conditions for up to 12 months can be expected. Note that two cable anchors are used, the first inside the inner barrier and the second outside the coating area. The first coating is applied, ensuring that it is void-free, especially around the cable and is then allowed to become tack-free. Prior to doubling back the degreased cables, a layer of the second coating is applied over the first to enable the folded-over leads to be properly embedded, reducing the likelihood of voids occurring in the

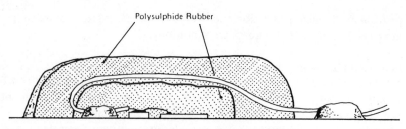

Polysulphide Rubber

FIG. 12.   Over-lap protection technique.

second coat around the leads. The use of a temporary spacer at the second anchor point will reduce the possibility of voids occurring, by keeping the leads separated. Again the application of a suitable paint to the installation will prolong useful life.

Long-term Techniques

Surround Technique (up to 5 years)
Neither the short-term nor the over-lap technique provides really substantial mechanical protection in addition to protection against moisture. In cases where mechanical protection is necessary (e.g. oil rigs) an overall metal surround and cable entry can be used. The 'surround technique', illustrated in Fig. 13, employs a double barrier system, in this case using two different materials, the inner a polysulphide rubber and the outer a polysulphide epoxy. This technique also allows a time gap between gauge laying and protection. After the gauge has been bonded to the structure, jumper leads are soldered to the gauge and bent at right-angles, the inner barrier is applied and a Tufnol bridge set into the coating before it has cured. The jumper leads are soldered to terminal pins in the bridge after the coating has cured. The installation can be left in this condition for some months before the outer barrier is applied, though this necessitates a thorough chemical cleaning of the inner coating and surrounding metal. Note that the cable/coating interface is extended and that the cable is supported in the tube, ensuring an even amount of coating material around the cable. The cable tube is raised up from the structure to reduce any stiffening effect. A single multicore cable is shown in Fig. 13, but for multicore cables the cable sheath is removed where the tube enters the case to permit adequate cable/coating length for each individual lead in case of damage to the outer insulation. If necessary, an additional protective metal cover can be used. This technique will adequately protect a gauge in full immersion conditions for up to five years. A suitable marine paint should be used to protect the coating/structure interface.

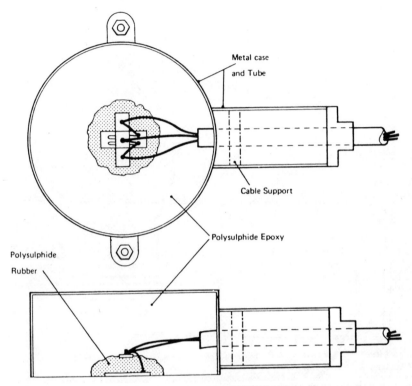

FIG. 13. Surround protection technique.

Multigauge Technique

In those situations where a long-term installation is required and where there are a number of gauge positions in close proximity to each other, there is often insufficient space for individual surrounds. In this case, each gauge can have a separate inner barrier within a common outer barrier, see Fig. 14. This again enables a time delay between gauge laying (protected by the inner barrier) and signal cable installation and the casting of the outer barrier. One major problem with such a technique is the high probability of failure of the complete area, because of water entering via an individual lead breakdown. To reduce this possibility, the insulation of each cable core is separated, permitting the encapsulant to provide a seal between the conductor and the insulation, thus making a water trap. Note that each lead is supported by a spacer to ensure all round protection. The temporary polythene mould (with rounded corners) is so arranged as to permit any

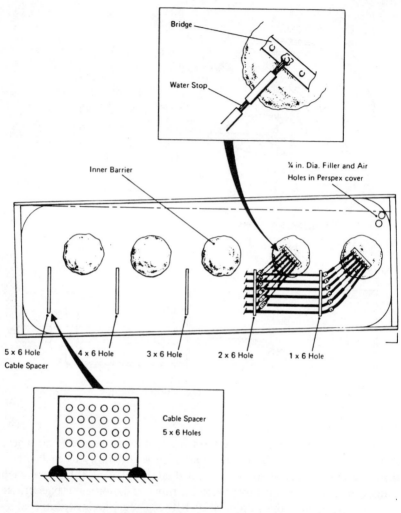

FIG. 14.   Multigauge protection technique.

trapped air to escape from the installation. When the polysulphide epoxy has cured, the mould and perspex cover are removed and, if necessary, metal protective covers are fitted.

Pre-encapsulation
As previously discussed, the hostility of the installation environment may limit the possibility of strain measurement in a particular situation. A

possible solution to this problem could be to remove such delicate operations as gauge handling, soldering and protection to better conditions, and one method of achieving this is to pre-encapsulate the gauge, as illustrated in Fig. 15. The pre-encapsulation technique has the following advantages:

    (a)   approximately 75% of the installation work is transferred to 'laboratory type' conditions, with improved quality control, reduction of installation costs and less interference with production;

    (b)   wiring of harness complete with gauges and glands, enables verification of water tight integrity of the system by immersion or pressure test, prior to installation;

    (c)   reduction of limitations on use of metal foil gauges due to hostile installation environment;

    (d)   gauging of areas which would be inaccessible with more conventional techniques.

There are, however, some disadvantages:

    (a)   exposed adhesive edge that may require additional protection;

    (b)   precise gauge alignment is more difficult;

    (c)   no access to gauge terminals for measurement, calibration or repair operations.

The encapsulation material can place a temperature limitation on the technique, and one material that has been successfully used, a semi-flexible polysulphide epoxy, has an upper limit of approximately 80 °C. However, the biggest limiting factor is the adhesive which is used to bond the encapsulated gauge to the surface. To exploit fully the pre-encapsulation technique, an adhesive which is not limited by climatic conditions would enable metal foil gauges to be laid in very hostile installation environments, e.g. pouring rain or even underwater. One particular polyester-based strain gauge adhesive can be mixed, applied and cured underwater. Using this technique, metal foil gauges have been successfully laid at a depth of 10 m in fresh water, reliable results being obtained. All the above techniques are discussed in greater detail in reference 13.

There are a number of environments in which it might be considered unnecessary to protect the gauge. However the pitfalls of this thinking are clearly demonstrated in Fig. 16. It might well have been considered that an unprotected open-faced foil which gave an RTG value of greater than 10^4 megohms when directly immersed in oil was perfectly satisfactory. However, the chlorinated EP oil used proved to be corrosive and the gauge

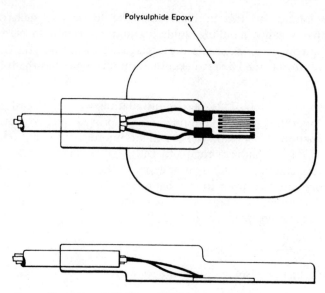

FIG. 15. Foil gauge encapsulation technique.

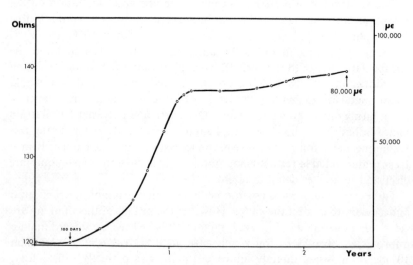

FIG. 16. Effects on open-faced gauge of EP oil.

foil etched away producing an apparent tensile strain reading of 80 000 microstrain after two years. It will be noted that the rate of etching was minimal for the first 100 days, increased rapidly for about one year, and then tailed off.

OBTAINING RELIABLE DATA

This section is not primarily concerned with errors in strain measurement (dealt with in detail in Chapter 5), but rather with ensuring that performance limits are not exceeded as a result of a hostile environment. There is invariably an area of doubt, and therefore uncertainty, when strain measurements are made in a hostile environment, because the effects of many environments on the measurement system are unpredictable. The effectiveness of, for example, the adhesive bond at high temperatures or high elongation is often unproved until the installation is subjected to the test environment, which is often too late for any corrective action. Laying a test gauge on a gauge factor bar in the actual installation environment, as proposed (Chapter 5), goes some way towards removing the uncertainty about any adverse effects of the *installation* environment on the strain transmission characteristics of the cured adhesive, but the effects of the *measurement* environment still remain unknown. One useful technique, which is an extension of the use of the gauge factor bar, is to subject the gauged bar to the actual or simulated measurement environment and observe the effects on the gauge output or gauge characteristics. In particular, this provides useful information when 'first cycle' data is required, as the response of the gauge on the gauge factor bar can often be related to actual test gauge response.

Invariably, the performance of the measurement system when initially subjected to the hostile environment is a good indication of short-term reliability, initial survival often indicating good prospects for short-term survival. Confidence in long-term survival is a more difficult proposition. Manufacturers are very reticent about this because there are so many user-dependent factors (which are outside their control), and commercial strain measurement services are unwilling to disseminate their knowledge. Therefore, it is up to the user to establish his own degree of confidence by realistic laboratory testing of samples during his development programme. This should be done as early as possible before the actual measurements are made, as an understanding of the reasons for any failure may prevent failure of the actual test gauges.

It is worth noting that failure of a particular type of installation in a

TABLE 15
STRAIN GAUGE CODES OF PRACTICES

Surface preparation	(a) Obtain a bright metal surface for *protective coating* as well as for adhesive.
	(b) Employ circular grinding and lapping to reduce direct tracking paths.
	(c) Use neutraliser last.
	(d) Check that last cotton bud used is as clean after use as before.
	(e) Do not test surface smoothness with fingers after chemical cleaning.
	(f) Ensure a minimum time gap between surface preparation and application of adhesive and coating.
Gauges	(a) Do not finger the gauge backing—use cleaned tweezers.
	(b) Do not bend the gauge backing/foil more than 25°.
	(c) Always record the Lot No., gauge factor, STC, K_t and ohmic value for each gauge.
	(d) If contaminated, clean gauge backing with *neutraliser* only (other solvents may contaminate).
	(e) Measure and record gauge resistance and RTG for each gauge, after laying.
Materials	(a) Use new packets of cotton buds, tissues, etc. for each job.
	(b) Destroy all packets of cotton buds, etc. that might be contaminated.
	(c) Use quality controlled, in-date metal conditioner and neutraliser.
	(d) *Never* place contaminated tissue, etc. on container mouth when moistening.
	(e) Always use cappable wash-bottles with identified contents.
	(f) *Never* use empty metal conditioner or neutraliser bottles for other solvents, etc.
Adhesives	(a) Use only recognised strain gauge adhesives, that are quality controlled and in-date.
	(b) Store away from ultraviolet light and refrigerate as necessary.
	(c) Check adhesive visually for inclusions, contaminations or discolouration.
	(d) Ensure containers are slightly *above* room temperature before opening, i.e. 'hand-warm'.
	(e) Read manufacturer's literature and observe safety precautions.
	(f) Mix two-part types thoroughly in well ventilated area. Avoid skin contact.
	(g) Apply correct pressure at right angles to the glue-line.
	(h) Before applying adhesive perform 'dummy run' with clamping.
	(j) Allow for heat conduction of large structures when applying heat.
	(k) Mix more adhesive than is actually required, to ensure efficient mixing.

Soldering

(a) Use correct type of solder, and temperature controlled iron with correct temperature bit.
(b) Pre-tin where possible.
(c) Do not use 'active' fluxes.
(d) Do not contaminate H T solder with L T solder left on bit-cleaning sponge.
(e) Where practicable secure items to be soldered before applying heat and solder.
(f) Unless essential, do not transfer solder to joint on the iron bit.
(g) Ensure that soldered joint has a low profile and no spikes.

Cable

(a) Use pressure-tested cable for full immersion conditions.
(b) Clean cable at coating entry point.
(c) Separate each core *before* entry into the coating.
(d) Ensure that no strands are damaged after stripping off insulation.
(e) Do not trap, stand on or otherwise roughly handle pressure-tested cables.
(f) Check compatibility of cable with any paints, etc., applied during manufacturing process.

Protective coating

(a) Use quality controlled, in-date materials only.
(b) Store away from ultraviolet light and refrigerate as necessary.
(c) Ensure containers are slightly *above* room temperature before opening, i.e. 'hand warm'.
(d) Read manufacturer's literature and observe safety precautions.
(e) Mix individual parts of two-part systems separately before combining, especially if dividing bulk materials.
(f) When mixing two-part systems use individual spatulas to avoid contamination.
(g) Mix two-part systems thoroughly in well ventilated area. Avoid skin contact.
(h) Always use a balance when dividing bulk materials.
(j) Do not use material which has to be scraped out of the container.
(k) Leave a sample coating in the job environment to check curing process and bond characteristics.
(l) Ensure material containers are properly closed and stored.
(m) Clean surface with neutraliser before applying coating.
(n) Measure and record gauge resistance after defluxing and before applying coating.
(o) Ensure efficient wetting of the surface by the coating.
(p) Ensure that there are no voids left under areas of cable entry.
(q) Do not use any material which liberates acetic acid on cure.

particular hostile environment by another user may have been caused by incorrect selection and application techniques rather than by the metal foil gauges having reached their limits of survival in the particular environment.

Assuming that selection of materials has resulted in a measurement system that will withstand the hostile environment, consideration must then be given to those environmental effects which will cause changes in bridge output other than those directly related to surface strain. The protection technique employed will offer protection from certain environmental effects, such as moisture ingress, but may not reduce the effects of others, e.g. temperature.

Although there is clearly a need to establish whether the indicated strain is due solely to strain induced gauge resistance changes, to the author's knowledge there is, unfortunately, no easy method of determining this. The main safeguards against non-strain induced resistance changes due to the effects of hostile environments are the correct selection and use of materials and adherence to good gauging practice. For example, in nuclear radiation fields, all materials exposed to the radiation must contain no organic compounds, and all electrical connections exposed to radiation should be welded.

Materials and techniques have been discussed in detail, but a third, and very important factor in establishing reliability, but one which is very difficult to quantify, is the attitude and integrity of the strain gauge operator. For example, two persons using identical materials, techniques, and within the same installation environment limitations, can produce strain measurement systems, one of which will survive very severe hostile measurement environments, and give credible data, while the other will fail as soon as the measurement environment becomes hostile, or even more serious, it will provide data which is unreliable.

It is often difficult to explain why an installation has failed, but in a number of cases within the author's experience, failure could be attributed, in part, to a particular person and his attitude to strain measurement. This clearly illustrates the problems in using an ideal laboratory measurement technique in a hostile environment. Adequate supervision at all stages of an installation by people qualified in detecting potential failure situations, including operator attitudes, will substantially reduce the risk of producing unreliable data.

To further reduce the possibility of human error affecting reliability, it is recommended that each organisation has an accepted 'Code of Practice' for strain measurement. Table 15 details such a code, which has been devised by the author as a result of observations and analysis of failed strain gauge installations. Although this code is relevant to all strain measurement systems, departures from it could be more crucial when measurements are being made in hostile environments.

CONCLUSIONS

Information on the behaviour of materials and structures in actual operating conditions is often required and it is only to be expected that strain measurements will be required in hostile environments. Strain measurement using metal foil gauges is ideally suited to 'laboratory type' conditions, and much of the available literature, user experience and thinking is related to these conditions; successful measurements in hostile environments require a different approach.

A number of interrelated factors require consideration, namely (a) the limitations of the installation environment on the selection of the strain gauge materials, and on the ability of the installer to use them successfully, (b) the effect of the measurement environment, including the time factor, on the ability of the measurement system to withstand the conditions and to provide reliable data for the duration of the test programme, and (c) the effects of any adverse premeasurement environment.

It is necessary to select materials that will survive in the hostile environment and to employ techniques that will adequately protect the gauge installation from the environment.

The effects of the two most commonly experienced hostile environments, temperature and fluid immersion, have been considered in detail as an aid to obtaining successful measurements in these conditions, and because they adequately illustrate the approach necessary to obtain reliable data in a number of less frequently encountered, yet equally hostile, environments.

Practical strain measurements are possible in a surprisingly large number of different environments, provided that those responsible for the selection of materials take into account the measurement environment, the premeasurement environment, and the installation environment, and have practical experience of the physical and mental effects of the environment on the operator, to ensure that everything humanly possible is done to enable the operator to realise his abilities to the full.

REFERENCES

1. POPLE, J., *Index of strain measurement papers*, (Unpublished).
2. *Strain Gauge Selection—Criteria, Procedures, Recommendations*, 1979, TN-132, Micro Measurements Division, Measurements Group Inc., Raleigh, USA.

3. *Bondable Printed Circuit Terminals*, 1978, A-138, Micro Measurements Division, Measurements Group Inc., Raleigh, USA.
4. *The Proper Use of Bondable Terminals in Strain Gauge Applications*, 1975, TT-127, Micro Measurements Division, Measurements Group Inc., Raleigh, USA.
5. 'Strain Gauge Solders', A-132, Measurements Group Inc., Raleigh, USA, Jan. 1978.
6. *Silver Soldering Techniques for Attachment of Leads to Strain Gauges*, 1975, TT-129, Micro Measurements Division, Measurements Group Inc., Raleigh, USA.
7. *Surface Preparation for Strain Gauge Bonding*, 1976, B-129, Micro Measurements Division, Measurements Group Inc., Raleigh, USA.
8. POPLE, J., *BSSM Strain Measurement Reference Book*, BSSM 281, Heaton Road, Newcastle upon Tyne, England, 1980.
9. PERRY, C. C. and LISSNER, H. R., *The Strain Gage Primer*, McGraw-Hill Book Company, New York, 1962.
10. *Soldering Techniques for Lead Attachment to Strain Gauges with Solder Dots*, 1974, TT-128, Micro Measurements Division, Measurements Group Inc., Raleigh, USA.
11. *Techniques for Bonding Leadwires to Surfaces Experiencing High Centrifugal Forces*, 1975, TT-132, Micro Measurements Division, Measurements Group Inc., Raleigh, USA.
12. *Lead Attachment Techniques for Obtaining Maximum Fatigue Life of Strain Gauges*, 1974, TT-130, Micro Measurements Division, Measurements Group Inc., Raleigh, USA.
13. BEARD, T. W., CRAWLEY, B. F. C. and POPLE, J., Strain gauging in a shipyard environment, *BSSM/RINA Conf.*, *Proc. from BSSM 281*, Heaton Road, Newcastle upon Tyne, England, Sept. 1975.
14. KING, N. E., *Failure of metal to epoxy resin bonds*, Thesis No. D6618/77 University of London, England.
15. DITTBENNER, G. R., New Techniques for Evaluating Strain Gage Waterproofing Compounds, *SESA*, Huntsville, Alabama, May, 1970.
16. DITTBENNER, G. R., Evaluating the quality of strain gauge waterproofing compounds (Letter to the Editor), *Strain*, **8** (No. 2), April 1972.
17. POPLE, J., The selection of coating materials for marine applications, *Presented BSSM Conf.*, City University, London, May 1973.
18. WEIR, R. M., The effects of earth leakage on ERS gauges, *Strain*, **5** (No. 2), 1969.

BIBLIOGRAPHY

1. BLOSS, R. L., Strain gages in hostile environments, *Workshop Session, SESA*, Albany, New York, May 1968.
2. TELINDE, J. C., Strain gages in cryogenic environments, *Exp. Mech.*, Sept. 1970.
3. KAUFMAN, A., Investigation of strain gages for use at cryogenic temperatures, *Exp. Mech.*, Aug. 1963.
4. BERTODO, R., Resistance strain gauges for the measurement of steady strains at temperatures above 650°C, *JSA*, **1** (No. 1), 1965.

5. SUMNER, G., Measurement of dynamic and static strain in small specimens at elevated temperatures, *JSA*, **9** (No. 2), 1974.
6. SHARPE, W. N., JR., Strain gages for long-term high-temperature strain measurements, *Exp. Mech.*, Dec. 1975.
7. BROAD, M. S., Measurements of static strain under rapid heating, *Strain*, **4** (No. 4), Oct. 1968.
8. FRISCH, J. and MORRIS, J. E., Strain measurements in tubes during rapid transient heating, *Exp. Mech.*, Aug. 1967.
9. DEAN III, M., Vulcanized-rubber protection for strain gages in a seawater environment, *Exp. Mech.*, Aug. 1977.
10. BEARD, T. W., CRAWLEY, B. F. and POPLE, J., Strain gauging in a shipbuilding environment, *Nav. Archit*, Jan. 1976.
11. CLARK, J. C., Strain measurement in the marine environment, *Presented at Mar. Env. Symp. SEE*, Dec. 1973.
12. SHOCK, R. N. and DUBA, A. G., Pressure effects on the response of foil strain gages, *Exp. Mech.*, Jan. 1973.
13. GERDEEN, J. C., Effects of pressure on small foil strain gages, *Exp. Mech.*, March 1963.
14. MILLIGAN, R. V., The gross hydrostatic pressure effects as related to foil and wire strain gages, *Exp. Mech.*, Feb. 1967.
15. BRACE, W. F., Effect of pressure on ERS gages, *Exp. Mech.*, July 1964.
16. FOSTER, C. G., Response of foil strain gages to hydraulic pressure, *Exp. Mech.*, Jan. 1977.
17. WERELDSMA, R., Stress measurements on a propeller blade of a 42 000 ton tanker on full scale, *Int Shipbld Progr.*, **2** (No. 113), Jan. 1964.
18. BOYLE, H. B., The strain gauge: an aid to the development of marine transport, *Strain*, **6** (No. 4), Oct. 1970.
19. BREWER, G. A., Operating stresses on SS Michigan propeller, *Exp. Mech.*, March 1972.
20. WEYMOUTH, L. J., Strain measurement in hostile environments, *Appl. Mech. Rev.*, **18** (No. 1), Jan. 1965.
21. WNUK, S., Progress in high temperature and radiation resistant strain gage development, *Exp. Mech.*, May 1965.
22. DAICHIK, M., *et al.*, The increase of strain gage measurement reliability under extreme conditions, *Exp. Mech.*, Nov. 1974.
23. STEIN, P. K., Strain gages in intense radiation fields, *Strain Gage Readings*, **1** (No. 2), Stein Engineering Services, Arizona, 1958.
24. SESA, Errors associated with strain measurement, *Proc. WRSGC Fall Mtg*, 1966.
25. WILLIS, D. L., Strain gages in high *G* environments, *Exp. Mech.*, Oct. 1969.
26. BARNES, S. G., Electrical noise, *Strain*, **8** (No. 4), Oct. 1972.
27. PEEKEL, C., Do we measure strain when we measure strain?, *Strain*, **8** (No. 3), July 1972.
28. JANZA, F. J., Pressure gages for severe blast environment, *ISA Preprint 19-1-Phymid-67*, Chicago, Illinois, Sept. 1967.
29. TOWLE, L. C., Measurement of resistance in hazardous atmospheres, *Int. Control and Instr. Conf.*, 1971.
30. MATHEWS, R. W., Intrinsically safe strain gages monitoring system, *ISA ASI 75241*, 247–9.

31. BEYER, F. R. and LEBOW, M. J., Long-term strain measurements in reinforced concrete, *SESA*, **XI** (No. 2), 1954.
32. ELMS, D. G. and JOHNSTONE, P. G., Strain gauge installation in a civil engineering site, *Strain*, **7** (No. 2), April 1971.
33. NOLTINGK, B. E., A review of techniques for telemetry from rotating parts, *CERL Conf. on Transmitting Signals from Rotating Plant*, Leatherhead, June 1970.
34. DOWLING, N. E., Performance of metal foil strain gages during large cyclic strains, *Exp. Mech.*, May 1977.
35. STRONG, J. T. and DANIELS, W. J. S., High elongation measurements in a hot water environment, *RD/B/N3387*, Nov. 1975.
36. BERTODO, R., High temperature high-elongation resistance strain gauges, *JSA*, **4** (No. 3), 1969.
37. PROKOPEC, M., Strain gauge measurements on plastics, *Strain*, **10** (No. 1), Jan. 1974.
38. TOOTH, A. S., Strain gauge measurements on glass, *Strain*, **11**, No. 1. Jan. 1975.
39. BUSH, A. J., Strain gages installations for adverse environments, *Exp. Mech.*, April 1969.
40. BATTEN, B. K., Five examples of the use of strain gauges in unusual environments, *Mar. Engrs. Rev.*, June 1972.

Chapter 4

INSTRUMENTATION

K. Scott

Welwyn Strain Measurement Ltd, Basingstoke, UK

and

A. Owens

Stress Engineering Services Ltd, Bath, UK

INTRODUCTION

The definition of strain gauge instrumentation is, for the purposes of this chapter, everything from the energising of the strain gauge to the production of test results in the required form.

The basic principles for accurate measurement were established many years ago. Modern technology offers better techniques, better resolution, the potential for higher accuracies, but primarily much higher speeds for data acquisition and analysis.

Almost all strain measurement systems can be broken down into the components shown in Fig. 1. The strain gauge itself is of paramount importance. The correct choice and proper installation is covered in other chapters and it is assumed here that the gauges perform correctly. A gauge can therefore be considered as a passive resistor which requires a power source. Changes in resistance caused by mechanical strain are measured in a bridge circuit which produces an out-of-balance voltage. This voltage needs to be amplified and displayed or stored, or both, after manipulating (or processing) it to represent the required units. This 'manipulation' may be by means of controls in the 'hardware' (analogue), e.g. gauge factor control. In a computer based system, all the manipulation may occur in the

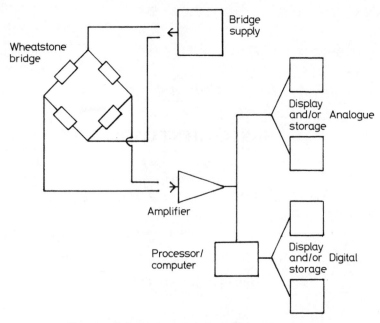

FIG. 1. Schematic—strain measurement system.

software (digital) either in the test programming, or in the data reduction and analysis, before or after storage.

For straightforward measurements of static strains from a small number of strain gauges, the commercially available self contained 'strain indicators' are robust, reliable, accurate, and almost foolproof. Similarly, for dynamic measurements from a few channels, the choice of suitable strain gauge amplifiers and recorders is quite straightforward.

For large test programmes, either static or dynamic, very elaborate systems can be built up from commercially available instrumentation and peripherals, or purchased as a purpose-built system.

WHEATSTONE BRIDGE

The Wheatstone bridge is most commonly used for converting the small change in the resistance of the strain gauge (or gauges) into a voltage suitable for amplification and processing. Reference is made later (see section 'Transmission and Noise') to the signal-to-noise ratio. It is at this early stage in the measurement chain that the most beneficial improvement in that ratio can be made by careful use of the bridge and its properties.

The principles given here are a basic look at the Wheatstone bridge and its use with strain gauges. For a full theoretical treatment, the reader is referred to the many text books available on the subject.[1,2]

Principles

Consider Fig. 2, in which R_1, R_2, R_3, R_4 are resistors. Assuming that the condition $R_1/R_4 = R_2/R_3$ is satisfied then the output voltage V_{out} will be zero, i.e. the bridge is balanced. A change in resistance in R_1 will unbalance the bridge and produce a voltage across the output terminals.

If a similar change, in both magnitude and polarity occurs in an adjacent arm of the bridge, say R_4, then the voltage V_{out} will remain at zero and the bridge will remain in a balanced condition.

If, in adjacent arms, resistance changes occur of equal magnitude but opposite polarity, then the voltage V_{out} will be twice that due to resistance changes in one arm. For strain gauge purposes the output equation for the bridge is

$$V_{out} = \frac{K\varepsilon N V_{in}}{4} \tag{1}$$

where K = gauge factor, V_{in} = bridge volts, ε = strain and N = number of active arms of the bridge (see following sections).

Two points to note are

(1)   The output is independent of gauge resistance.
(2)   Under certain conditions the Wheatstone bridge is non-linear, a second order term having been neglected in the equation given (see 'Wheatstone Bridge Linearity').

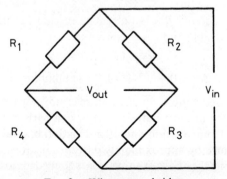

FIG. 2.   Wheatstone bridge.

The output voltage, V_{out}, can be directly measured and the required strain computed. This is known as the 'direct readout' or 'deflection' use of the bridge. The output can alternatively be restored to zero by the use of a previously calibrated variable resistance in another arm of the bridge. This is known as the 'null-balance' use of the bridge and further reference is made to it in the section 'Reference Bridge'. Suffice it to say at the moment that it is not now used as widely as the deflection mode.

The output voltage, V_{out}, is also dependent on the resistance or impedance of the circuit that is connected across the output terminals of the bridge. In early instruments this could have been a low impedance galvanometer or meter and its effect on the output voltage would have been severe. It would have been more usual to quote the output from the bridge in terms of current rather than voltage. With modern amplifiers of extremely high impedances there is little or no 'loading' effect on the bridge, i.e. the current consumed by the external circuit is negligible.

Input Configurations
Quarter Bridge Operation
When a single gauge is used at the measurement point, with perhaps resistors within the strain indicator completing the bridge, then that is termed 'quarter bridge' operation. For stress analysis it is now a most widely accepted technique and has been made possible by the self-temperature compensated foil strain gauge. It must be remembered that the lead wires connecting the gauge into the bridge form part of the measurement circuit and must be treated with care. To this end the three lead wire system (Fig. 3) is recommended. It is interesting to look at the order of resistance changes we are dealing with in the strain gauge. From

$$\Delta R = K \times R \times \varepsilon \qquad (2)$$

where K = gauge factor, R = initial gauge resistance and ε = strain and using $K = 2$, $R = 120\,\Omega$ and $\varepsilon = 1\,\mu\varepsilon$ we find

$$\Delta R = 2 \times 120 \times 1 \times 10^{-6}$$
$$\Delta R = 0.000\,24\,\Omega \text{ or } 0.24\,\text{milliohms}$$

We are therefore trying to detect and measure resistance changes that would be neglected in most industries. For comparison a two-conductor copper cable of $1\,\text{mm}^2$ section and $10\,\text{m}$ length (i.e. total of $20\,\text{m}$) would change in resistance by approximately 0.0128 ohms for a $10\,°\text{C}$ change in temperature. Using the above example this is equivalent to a shift of $54\,\mu\varepsilon$ in the gauge. Depending on the mechanical strain being measured this could

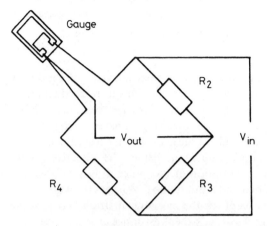

FIG. 3. Quarter bridge, three lead system.

be a considerable error. The three lead wire system offers the following advantages:

(1) The resistance of the lead wires, irrespective of length does not unbalance the bridge because there are similar lead wires in adjacent arms of the bridge and their effects will cancel out.

(2) Resistance changes of the lead wires produced by temperature changes are cancelled out in the bridge.

(3) Desensitisation effects of the lead wires are halved.

The lead wires should be as near identical as possible and perhaps twisted so that they each see the same temperature environment.

Applying the output equation (eqn. (1)) for the quarter bridge case, using $K = 2$, $V_{in} = 1$ volt, $\varepsilon = 1\ \mu\varepsilon$ and $N = 1$ (active arm) then

$$V_{out} = \frac{2 \times 1 \times 10^{-6} \times 1 \times 1}{4} \quad \text{volts}$$

$$V_{out} = 0 \cdot 5\ \text{microvolts}$$

This illustrates the minimum order of signal levels we are dealing with and hence the care and attention to detail of both the installation and instrumentation in order to resolve to that very low level.

Note, with both the quarter bridge three lead system and the half bridge (see following section) there is a choice as to whether the third lead goes in the supply or signal line. Commercial manufacturers seem to favour the latter as shown in Fig. 3. Variations do occur, however, even within one manufacturer's range. Further references are made in Chapter 5.

Half Bridge

When two gauges are used in adjacent arms of the bridge (Fig. 4) it is known as a half bridge system. For this to work then one gauge must see tension and the other compression, i.e. one increasing and the other decreasing in resistance, or one gauge must see zero mechanical strain. For example, when a single gauge is used to measure the mechanical strain at a temperature either above or below the effective range of the self-temperature compensation of the gauge, it is advisable to use a 'dummy' gauge, positioned and mounted so that it sees the whole environment of the 'active' gauge except for the mechanical strain being measured. With the two gauges wired as Fig. 4 then resistance changes of the same sign caused

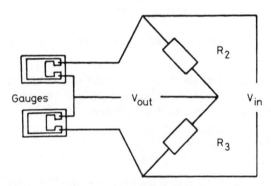

FIG. 4.   Half bridge system.

by temperature changes will not unbalance the bridge. Note that, however, the use of the half bridge in any form does not ensure complete and automatic temperature compensation. It is difficult to maintain perfect symmetry and it relies on the two gauges having well-matched temperature/resistance characteristics.

Two examples of where a half bridge system might be used are:

(1)   Figure 5, a cantilever beam with one gauge on top and one directly underneath wired as Fig. 4. If the output is measured and eqn. (1) (see 'Principles') rearranged to compute the strain, then with $N = 2$, the average of the two surface strains will be recorded, with $N = 1$ the sum of the two surface strains will be recorded.

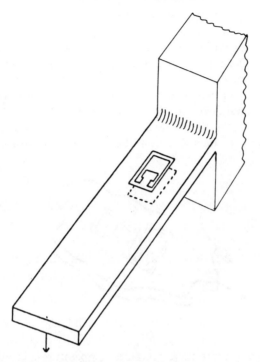

Fig. 5.   Cantilever beam using half bridge.

(2)   Figure 6, a specimen in tension with two gauges, one arranged
       along the principal axis and the other at right angles, again wired as
       Fig. 4. The axial gauge will see the tensile strain and the gauge at
       right angles will see a compressive strain, thus satisfying the
       requirements for a half bridge. The compressive strain will be a
       proportion of the tensile strain depending on the Poisson ratio (v)
       for the specimen material. This will have to be taken into account if
       the axial surface strain is to be computed using $N = 1 + v$. $N = 1$
       will give the sum of the two strains.

Full Bridge
The full bridge, using gauges as all four arms of the Wheatstone bridge, is a
logical extension of the half bridge and can be used to further increase the
sensitivity of a measuring system. Putting two gauges on each side of a
beam instead of one, gives a value of $N = 4$ i.e. the output is four times that

of a single gauge quarter bridge installation, with possibly improved temperature compensation, and the cancellation of unwanted signals. In a full bridge the value of N can be 1, 1·3, 2, 2·6 or 4, depending on the gauge arrangements. The greatest advantage of the full bridge is that all the lead wires from the measuring point to the instrumentation, including plugs, connectors, and sliprings if used, are *outside* the measuring circuit and contribute minimal errors to the system (see Fig. 7).

The full bridge is therefore often used for installations where the measuring point is a long way from the reading instruments, particularly

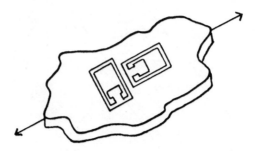

FIG. 6.   Tensile specimen, half bridge.

for 'on-site' work where there are large temperature and other environmental variations. The full bridge is also used for measurements on rotating machinery requiring the use of sliprings. As the sliprings are not within the measuring bridge, contact resistance and contact resistance variations are much less critical. However, the most common use of the full bridge is in transducers, which are calibrated directly to read in engineering units other than strain, and the value of N is accounted for in that calibration. Examples are: bending beam; shear beams; torque meters and diaphragm pressure gauges (all with $N = 4$) and tension and compression cells ($N = 2·6$).

Bridge Supply

So far the Wheatstone bridge has been discussed as if powered by a direct current, constant voltage supply. There are other means of supplying bridge power, a complete discussion of which is beyond the scope of this book. Neither is it possible to say that one system is generally better than another, they all have specific advantages or disadvantages which might or might not lend them to particular tasks.

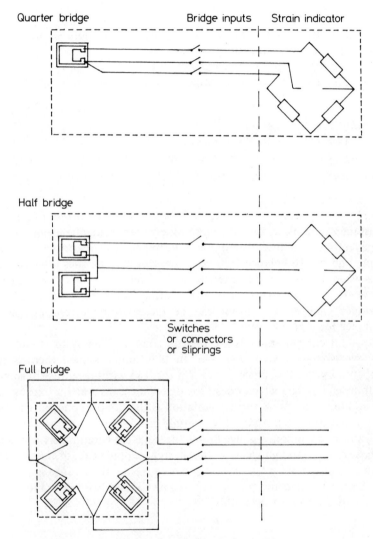

FIG. 7. Bridge configurations—measuring circuit. All items within the dotted boundaries are in the bridge (measuring) circuit and will directly affect the measurement.

Advantages of d.c. over a.c.
  (1)   Simplest approach in terms of circuitry involved.
  (2)   Wide frequency response.
  (3)   Sensitivity of system i.e. the amplifier gain is more stable.
  (4)   Better system linearity.
  (5)   No cable capacitive or inductive effects.
  (6)   Shunt calibration and bridge balance circuitry are simpler.

Advantages of a.c. over d.c.
  (1)   Less sensitive to electromagnetic interference.
  (2)   Does not measure thermocouple voltages.
  (3)   Any amplifier zero drift does not affect system zero.

Direct Current: Constant Voltage (CV) and Constant Current (CC)
Constant voltage is perhaps the more widely used technique in equipment
for general stress analysis. It requires less complicated circuitry for both the
supply and bridge balance facilities. Constant current does offer advantages when dealing with large resistance changes in that any non-linearity
effects are reduced. Also, because the output is a direct function of the
resistance change, the output can be increased by increasing gauge
resistance for a given strain level.

   Constant current supplies are used in some logging systems and the
principle is illustrated in Fig. 8(a). In that simple system there is no
compensation for resistance changes in the lead wires. Figure 8(b) shows
the four lead system which does provide compensation at the expense of
two additional leads. A further variation is the double constant current
system as in Fig. 8(c). Again this can only give lead wire compensation if
the leads to the 'dummy' gauge follow the same temperature environment
as those of the active gauge. Constant current supplies can, of course, also
be used with the Wheatstone bridge circuit. (Note that transducers that
have been calibrated and compensated using CV may lose both calibration
and compensation if used with CC.)

Alternating Current: Sine Wave or Pulsed Waveforms
Cable capacitance effects are reduced with pulsed waveforms. Also the
RMS (heating effect) is lower using a pulsed system than with a continuous
sine wave of similar amplitude. (Note: a square wave is a form of pulsed
wave shape with a 1:1 mark–space ratio.)

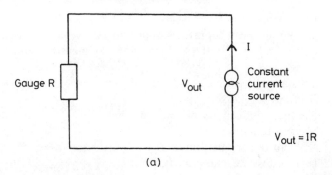

(a)

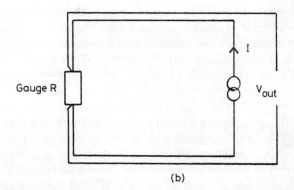

(b)

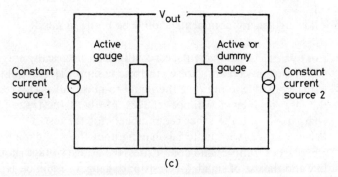

(c)

FIG. 8. (a) Simple constant current circuit. (b) Four lead circuit. (c) Double constant current circuit.

Bridge Energisation Level

Irrespective of the type of supply, a suitable power level has to be chosen to take account of all the variables which affect the measurement:

(a) Resistance of gauge.
(b) Grid area of gauge.
(c) Thermal conductivity of test specimen.
(d) Mass of test specimen.
(e) Ambient test temperature.
(f) Type of test, i.e. static or dynamic.
(g) Accuracy requirements of test.
(h) Whether long- or short-term measurement.

Some commercial strain indicators, usually those intended for static work, have a fixed level which should have been chosen to be sufficiently low to allow small gauges to be used on poor heat sinks. Other commercial instruments have a variable supply usually in the range, 0·5 to 12 volts.

A strain gauge is essentially a resistance which, when a current is passed through it, has to dissipate the heat generated. Most of that heat is dissipated by conduction through the material to which the gauge is bonded. If the heat is not conducted away at a sufficient rate then the gauge temperature rises and its resistance changes. The effect then is a continuing drift on the output of the strain indicator. If the requirement is primarily to measure dynamic strain, then some 'drift' may be acceptable but in static work good stability is of prime importance. With all the variables involved it is not possible to stipulate gauge excitation levels. However, most gauge manufacturers supply data which will give a good indication of acceptable power densities.[3]

The following general comments should be borne in mind:

(1) For dynamic measurements, especially in an electrically noisy area (see 'Transmission and Noise') the bridge supply level should be as high as possible to improve the signal-to-noise ratio. This requires the use of high resistance gauges of the largest size (area) compatible with the other requirements for the test.

(2) When working with materials having poor thermal conductivity, e.g. plastics, rubbers and composites, the bridge voltage should be low and the use of small, low resistance gauges should be avoided. The physical size might be dictated by other considerations, i.e. the size of the component itself and strain gradients.

Wheatstone Bridge Linearity

As mentioned earlier, the Wheatstone bridge is non-linear under certain conditions (see also Chapter 5) generally when there are non-symmetrical resistance changes within the bridge and when large resistance changes are involved. The quarter bridge is perhaps the most widely used example. For most measurements on metals within the elastic range where the changes in resistance are small, this non-linearity is negligible. As a rule-of-thumb the non-linearity is equal to the percentage strain, i.e. 0.1% strain $(1000\,\mu\varepsilon)$ would give 0.1% non-linearity. This assumes a constant voltage supply and, as stated earlier, the effect can be reduced with a constant current supply.[4]

Bridge Balance

In most strain measurement systems there is the provision for initially balancing the bridge to compensate for resistance tolerances of the gauges and lead wires. Some means of establishing this arbitrary zero at the start of a test is needed for two reasons:

(a) to enable the strains to be read directly without having to add or subtract the initial offsets;

(b) to ensure that the linear range of the amplifier is used to its full extent for the required signal and not taken up by the initial offsets.

The most common method is to use a potentiometer and a fixed 'limit resistor' as in Fig. 9. The important value is that of the 'limit resistor' and this is chosen by the manufacturer to give a reasonable range of balance (e.g. $\pm 5000\,\mu\varepsilon$) without introducing unacceptable errors. Obviously the quality of the gauge installation is important in both choosing gauges that

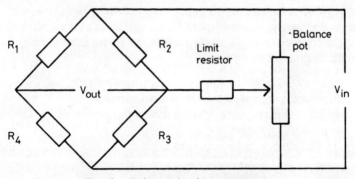

Fig. 9.   Balance circuit arrangement.

have a close resistive tolerance and ensuring that they are installed with minimum change in their resistance. Any balance system which directly acts on the bridge circuit itself will introduce errors. The magnitude of the errors should be quite small (less than 1 %) and the equipment manufacturer should be able to supply correction formulae if required.

A technique which does not introduce errors is the voltage-injection or auto-balance circuit. In this a voltage of equal magnitude to the initial offset is developed and applied to the amplifier to restore the output to zero. The applied voltage is digitised and held in a non-volatile memory which allows power to be removed from the unit without loss of balance.

It is also possible, with computer-based equipment, to take account of initial offsets in the software. The value of the offset would be stored and then added to, or subtracted from, subsequent readings as the polarity dictates.

Reference Bridge

So far we have looked at the bridge in its deflection mode, i.e. directly measuring the unbalance voltage to determine strain. Another operational mode is termed 'null-balance' in which the unbalance voltage is restored to zero by a calibrated variable resistance in another arm of the bridge. The meter or indicator then only determines the zero or null position and need not be of great accuracy. This technique is now not in general use because it can only be used with quarter and half bridge systems, but a variation of it, the reference bridge, forms the basis of some commercial strain indicators. The principle is illustrated in Fig. 10.

The gauge or gauges are wired into the 'active' bridge. The reference bridge is adjusted to give a voltage equivalent to the offset of the gauges, this being indicated by a simple zero meter indication. The reference bridge control can be calibrated to provide a direct readout in microstrain. The bridges are in fact powered by a 1 kHz square wave (1·5 V r.m.s.) and so additional circuitry is required to provide the balance indication.

The advantages of this system are:

(1) Sensitivity is not a function of bridge supply which makes it suitable for battery operated, portable instruments.
(2) Being a square wave system it is not susceptible to lead wire capacitance effects.
(3) The readout can be a rugged mechanical system, again lending itself to portable use.

It is, of course, primarily designed for static work but does possess a

1 kHz Square wave

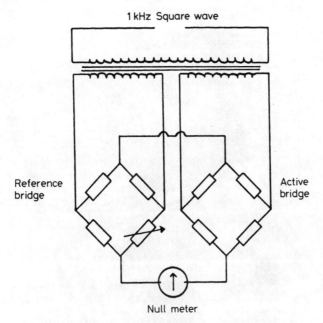

Reference bridge

Active bridge

Null meter

FIG. 10. Principles of reference bridge system.

dynamic capacity (feeding an oscilloscope, for example) limited to 20 % of the carrier frequency, i.e. 50 Hz. Figure 11 gives an example of a commercial unit.

Shunt Calibration
Shunt calibration is a means of providing an electrical simulation of a gauge's change of resistance. It can, therefore, be used to give an electrical calibration of the complete system from gauge through amplifier to readout, perhaps to establish that the correct display range has been chosen. It is important to note that it does not verify the gauge's mechanical performance nor that of the adhesive. The only way to check the sensitivity of a strain gauge is to apply a known strain to it, and this is very seldom possible in stress analysis applications. However, the quality of the gauge installation can be checked by applying any load or force which can be removed completely to check the zero return of the gauge installation. If the load can be repeated accurately, then the repeatability of the gauge installation can also be checked.

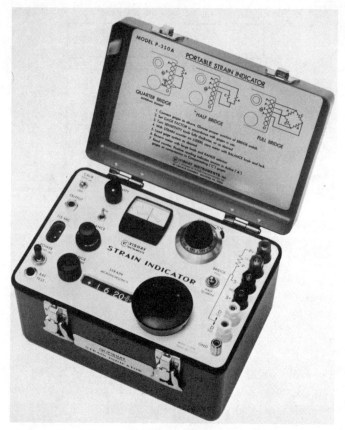

FIG. 11.   A portable battery operated strain indicator. Courtesy of Measurements
Group, Vishay.

Principles

The principle of shunt calibration is illustrated in Fig. 12 where R_{cal} is put
in parallel with one arm of the bridge (R_1) to simulate a resistance change
due to strain. Using eqn. (2) and simple resistors-in-parallel theory, then
the simulated strain ($\mu\varepsilon_{cal}$) can be derived from

$$\mu\varepsilon_{cal} = \frac{R_1}{K'(R_{cal} + R_1)} \times 10^6 \tag{3}$$

where K' is the effective gauge factor of the active gauge. The calibration
resistor can be used with any arm of the bridge, active gauge or bridge

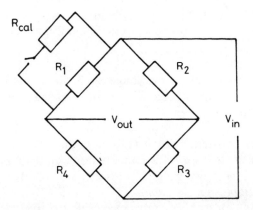

Fig. 12.   Principles of shunt calibration.

completion resistor. The two most widely used techniques are detailed
below with their advantages and disadvantages.

Shunting the Dummy Resistor (Fig. 13)
This system is used only with quarter bridges and has the advantage of
automatically correcting for lead wire resistance, because the shunted arm
is the same resistance as the active arm. The shunt resistor value is chosen
for the correct gauge resistance (e.g. for 120 or 350 ohms). If the lead wire

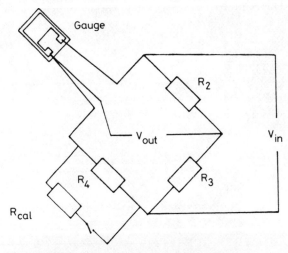

Fig. 13.   Shunting the dummy resistor.

resistance in series with the gauge is significant, then as it is also in series with the dummy resistor, the effect of the shunt will be correctly modified and the instrument readout can still be set at the normal strain value on the calibration switch.

Only tensile strain in the active gauge can be simulated by shunting the dummy resistor. In this instance $K' = K$, the package gauge factor of the active gauge (eqn. (3)).

Shunting the Internal Half Bridge (Fig. 14)
This system can be used with either quarter or half bridges and is independent of the active gauge(s') resistance. It will also simulate both tension and compression. K' will not be equal to K and a correction will have to be made for lead wire resistance using the equation

$$K' = K \frac{R_g}{R_g + R_L} \tag{4}$$

where K = package gauge factor of active gauge, R_g = resistance of gauge, and R_L = resistance of lead wire(s) in series with the active gauge. Note that with the quarter bridge three lead system and half bridge, R_L is the resistance of *one* lead wire.

If, as in many instruments, the internal half bridge that is used for calibration is also used for bridge balance using the potentiometer method (see 'Bridge Balance') then strictly speaking, a further correction will be necessary. This is due to the presence of the potentiometer and limit resistor

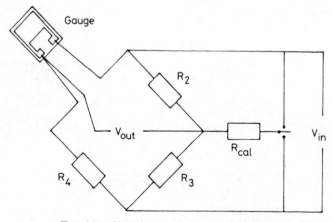

Fig. 14.   Shunting the internal half bridge.

affecting the resistance of the shunted arms. The correction value depends on the position of the balance potentiometer which is, of course, a function of the initial out-of-balance of the active gauge(s). It is normally of a low order, but if corrections need to be made then the equipment supplier should be able to provide suitable formulae.

AMPLIFIERS

For our purposes an amplifier is defined as a two-input device which can consist of a number of stages. A strain gauge amplifier is assumed to include bridge completion resistors suitable for quarter (three lead) and half bridge use, balance facilities, and a power supply for the bridge. It would also be designed with dynamic use in mind. A strain indicator would further include some means, such as a digital voltmeter, of displaying the signal in microstrain and would be used primarily for static measurements.

There have been major changes in the design and specification of amplifiers in recent years. Early strain indicators did not use any amplification, relying on a sensitive optical galvanometer to display the bridge output. It was very difficult to build stable d.c. amplifiers using valves and hence a.c. bridges and amplifiers were more common.

It is also true that with early indicators, electrical noise was not so much a problem because of their limited frequency bandwidths. Also of course there were fewer motors, generators etc. around to create electrical interference and be a profound nuisance when making any low level measurement such as strain.

Modern electronic techniques have brought about a proliferation of amplifiers, some of which are suitable for strain measurement and some of which are not. The type of amplifier is usually very much dictated by the nature of the task, i.e. static or dynamic measurements.

A static signal is one that does not intentionally change in magnitude to any extent with time, whereas a dynamic signal will vary rapidly with time. There is a 'grey' area in between, but most strain measurements can be split into these two groups. It is of prime importance that this is recognised at an early stage in project planning as it can determine the type of gauge used, the wiring, instrument and method of data processing.

With static measurements it is usually possible to use only one amplifier and, if more than one measurement point is needed, then a number of gauges can be switched, feeding one at a time, to the amplifier. In dynamic work correlation between different measurement points is usually required which means simultaneous amplification and recording and hence one amplifier per measurement point.

In a complete strain indicator system the limitations are in two areas, the design of the amplifier itself and the means of display. In a static system the amplifier might well have a reasonable dynamic response but a digital display would not be suitable for dynamic work.

Commercially made equipment is usually designed with one particular application in mind i.e. static or dynamic. The prime requirement of a static system is stability, i.e. with a fixed input the amplifier output will remain constant with time, and over the normal working temperature. For some applications such as long-term creep in materials, then good stability might be required for years. This is not easy to achieve and is very much dependent on the amplifier design, mechanically and thermally as well as in an electrical sense. For dynamic work, an amplifier will be designed to have a wide frequency response and low internally generated noise. There is, however, a degree of incompatibility between these two features, i.e. noise is directly proportional to bandwidth. With amplifier gains of the order of 10 000 being available, it is important that steps are taken to minimise internal pick-up from, say, a mains power supply.

The means of display and/or recording of results will differ depending on whether the work is static or dynamic. The various peripheral devices are discussed under 'Display' and 'Storage of Data' but their choice is dependent upon the amplifier. For static work a high impedance or voltage output is usual to feed, for example, a digital or analogue display. The range of devices used for dynamic work is more varied and the amplifier should be able to interface with both high and low impedance inputs, i.e. provide both a voltage and a current output. The latter would be used for ultraviolet (UV) recording equipment.

An example of a commercial dynamic strain gauge amplifier is shown in Fig. 15. This has a frequency response up to 25 kHz (± 0.5 dB) and a gain range from unity to 11 000. It has built in filters and an automatic bridge balance circuit which removes the influence of the potentiometer method on the input circuit.

Nowadays such extensive use is made of strain measurement equipment and the demands placed upon it by various authorities are so rigorous that confidence in an amplifier's correct performance is important. Shunt calibration does give reasonable validation for particular tasks but it is useful to have an accurate standard to which different amplifiers used within a laboratory can be referred. A suitable device is available commercially. This instrument presents a true Wheatstone bridge for both

FIG. 15. A commercial dynamic amplifier and bridge conditioning system. Courtesy Measurements Group, Vishay.

120 and 350 ohm gauge resistances. It can be switched to give offsets equivalent to $100 \mu\varepsilon$ steps up to $\pm 100\,000 \mu\varepsilon$ in half and full bridge and $\pm 50\,000 \mu\varepsilon$ for quarter bridge with an accuracy of 0.025%.

TRANSMISSION AND NOISE

The transmission or conveyance of signals, specifically low level ones, is a necessary part of any measurement chain. The prime requirement of any transmission system is to carry faithfully the required information without allowing external influences to degrade the validity of that information. The signal route might involve quite sophisticated equipment such as long-range telemetry links, but inevitably somewhere a simple lead wire would be used to connect the strain gauge or transducer to the system. It is sometimes the simple lead wire that is the most susceptible part of the system and the area where perhaps some attention would yield the biggest improvement.

Success or failure in measurement can be illustrated by one expression, the signal-to-noise ratio. The greater the ratio then the more successful the measurement will be. Looked at like this, there are only two ways to improve the situation, increase the signal level or decrease the noise. We have already looked at techniques for increasing the signal and these can be summarised as follows:

(a) Use a half or full bridge system wherever conditions permit.
(b) Use the highest bridge supply voltage feasible.
(c) Use a higher resistance gauge to allow a higher bridge voltage to be used.

The techniques for noise reduction are less well defined and will be looked at in connection with particular transmission systems. Noise, defined as an unwanted signal that interferes with the validity of the required signal, can be split into two groups—spurious resistance changes and induced voltages.

Spurious Resistance Changes

These are mainly a problem when they occur within the bridge circuit itself, i.e. the lead wires of a quarter or half bridge installation, or the interbridge wiring of a full bridge. Remembering that we are in effect using very small resistance changes to measure strain, then any unwanted resistance change can produce large errors in both static and dynamic work.

Temperature Effects

The effect of temperature changes on lead wires in the quarter bridge case has already been looked at under 'Quarter Bridge Operation'. There it was shown that the three lead wire system will give good compensation providing certain conditions are met, i.e. the lead wires must be physically similar and see the same temperature environment. Note that under conditions of radiant heating the popular colour coding of wires can lead to temperature differentials caused by different heat absorption rates, e.g. black and white. These differentials can be quite significant in long cable runs. With the half bridge circuit the lead wires provide compensation as in the quarter bridge three lead system. The signal and power leads in the full bridge case are generally immune from this problem as they are outside the measuring part of the bridge. It would be possible, however, under severe conditions, for example at temperatures which require the use of high resistance nickel–chrome lead wire, for resistance changes due to temperature to cause variations in the supply voltage at the bridge and hence a change in the sensitivity of the system (see Chapter 5). This could be corrected for by using a six lead wiring system to the bridge. The additional two leads sense the actual voltage at the bridge and form part of a control loop which automatically adjusts the supply at the instrument to take account of variations. The instrument must obviously be fitted with the remote sense facility, which can also be used when supply voltage losses are incurred due to long leads. The effect of temperature can also be seen when poor quality resistors are used for bridge completion purposes. Bridge completion resistors should have a very low temperature coefficient ($1 \, \mathrm{ppm}/^{\circ}\mathrm{C}$) as well as a close resistance tolerance ($\pm 0 \cdot 01 \, \%$). A common technique in commercial strain indicators is to use gauges bonded to a metal plate which is mounted in a strain-free condition in the instrument, instead of conventional commercially available resistors.

Connectors

Invariably connections of some description will have to be made in all strain measurement exercises. As with resistance changes due to temperature, contact resistance changes will be of most consequence when they occur within the measuring part of the bridge. There is one great distinction, however, with contact resistance problems in that they are random occurrences and cannot be compensated for by, for example, the three lead system. In order of preference, joints can be soldered (or welded), screw terminals used, gold plated spring loaded terminals used or other connectors used (see Chapter 5). If permanent joints cannot be used it is important that good quality connectors are used and that the mating surfaces are kept in a very clean condition.

Once a connection has been made, then the contact resistance should not change and it can be balanced out. If the connector is subject to vibration, then contact resistance changes could occur and give erroneous results. Depending on the type of connector, repeated making and breaking can cause the resistance to decrease due to the self-cleaning and polishing action of the contacts. Eventually the plating will be eroded causing an increase in resistance (see also Chapter 5). Figures quoted for contact resistance vary according to the type of connector and materials used. Typically they would be between $2\,m\Omega$ and $15\,m\Omega$. As stated earlier, this would be balanced out. It is *variation* in contact resistance which is the problem, and once again, it is largely eliminated if the contacts are outside the measuring part of the bridge.

Switching/Scanning

Defining switching and scanning first; switching implies manual operation such as in a static system where a number of gauges are required to be monitored by a single amplifier or indicator. Scanning is the automatic operation of a similar function.

Comments made for connectors could well apply to switches used in strain gauge circuits. Switching within the bridge, i.e. quarter or half bridge circuits can only be achieved with the highest quality components. Repeatability of switching or scanning should be of the order of $1\,\mu\varepsilon$ which implies consistency in contact resistance of $0.24\,m\Omega$. This assumes a 120 ohm gauge resistance and again the effect can be reduced by using a higher resistance gauge.

Switching a full bridge will not usually be a problem as the contacts will be in a high impedance line, i.e. the input to an amplifier (approx. $10\,M\Omega$). Also switching or scanning after the amplifier should not present a problem.

Methods of dealing with data have changed rapidly in recent years allowing much faster acquisition and processing. This has meant that mechanical and electromechanical scanning devices could be the limiting factor in logging systems. Use is now being made of solid state multiplex circuits that are capable of very fast switching speeds under computer/processor control.

Sliprings

In many measurement systems it is necessary to obtain data from rotating components. There are two techniques for this purpose, short-range telemetry which will be considered later, and sliprings.

The latter have been used for many years and are still very useful in a wide range of applications. In order to minimise contact resistance, precious metals are generally used for the mating surfaces and in some instances air-cooling is recommended for more stringent applications. Rotational speeds in excess of 20 000 r.p.m. are possible and a standard temperature range of $-40\,°C$ to $+110\,°C$ is achievable. Slipring manufacturers should be consulted for more information and advice for specific applications.

Even with good quality sliprings, the variation of contact resistance makes all but full bridge operation very difficult. In most applications, such as torque measurement, this is no great hardship as a full bridge configuration is advisable for a number of other reasons, such as the minimisation of bending effects and increase in signal.

As with all other contact resistance problems, the 'noise' generated is a direct function of the current passing across the contacts and this can be reduced by increasing the gauge resistance. The use of $1000\,\Omega$ gauges is common in torque applications and most gauge manufacturers can supply patterns specifically for that purpose.

There are other techniques for reducing the current through the sliprings.[5,6]

Strain in Lead Wires
The lead wires are part of the measuring system and if subjected to strain will change in resistance, acting in the same way as a strain gauge. In some installations it is necessary to attach the lead wires to the structure under test. This should be done carefully, leaving small expansion loops and routeing the wires through low strain areas where possible.

To a certain extent the three lead wire system with a quarter bridge will reduce the problem but this relies on having similar effects in the three leads, which perhaps cannot be guaranteed under all circumstances.

Induced Voltages
Noise due to induced voltages comes mainly from electromagnetic fields which affect both gauges and lead wires. It is a problem mainly in dynamic applications where a wide frequency response is required from the system.

Effects on Lead Wires
Electromagnetic fields simply consist of two different field components, electric and magnetic. *Electric fields* are created in the

presence of a voltage, an example being fluorescent lights where a relatively high voltage is developed to start the discharge. A charged field will be built up around the source and charges being transferred to local conductors will cause a noise voltage to be fed to the strain gauge amplifier. The method of minimising this pick-up is to provide an electrically earthed screen between the source and the signal conductors which will 'collect' the charges and route them to earth. Screened cables are therefore recommended to reduce the effect of electric fields, the screening efficiency being a function of its coverage, e.g. conductive plastic screens being better than copper braid. The screen should be earthed at one point only and usually the amplifier end is the more convenient. It is quite possible to have different earth potentials (voltages) in a mechanical structure which would cause a current to flow through the screen creating a further noise source if it is earthed in two places. Most commercial strain gauge amplifiers intended for dynamic work have the facility included for earthing the cable screen.

Magnetic fields are created by a current flow to devices such as transformers, motors etc. Any conductors which pass through the magnetic field will have a voltage induced which will be faithfully amplified to produce a noisy signal. The analogy here is that of a transformer in an electrical circuit. There has to be 'relative motion' between the field and the conductor so either the field is produced by an alternating current (a.c.) or the field is stationary (d.c.) and the conductor is moving. An example of the latter is any conductor moving within the earth's magnetic field, the voltage induced is very small and in most cases is neglected.

It has been shown[7] that screening against magnetic fields with common materials is not effective, except with a 1″ rigid steel conduit. In this study a special tape having a high permeability worked well but was ruled out on cost grounds.

The other technique used for reducing the effects of magnetic fields is to arrange for self-cancellation of the induced voltages in the lead wires themselves. This is achieved by twisting the leads such that voltage or current induced in one loop is cancelled by that of the adjacent loop.

It was shown that about 1 cm is the optimum pitch for the twist for commercially produced wires. Other noise voltages can be picked up by cables, perhaps due to poor insulation resistance. These generally will create similar voltages (in phase and of the same magnitude) at the input to the amplifier. Most strain gauge amplifiers employ differential inputs which will reject these common-mode voltages, the rejection efficiency is called the common-mode rejection ratio. This is usually quoted by the manufacturer as, for example, 100 dB at 50 Hz. Most of the noise in the UK is around the 50 Hz frequency and the rejection ratio decreases at higher frequencies.

It is therefore advisable for dynamic work to use a twisted screened cable for any low level measurement and especially for strain gauge work. Two other points are worth remembering. Careful cable routeing can assist greatly in reducing interference. Do *not* coil up the excess length in a cable. This is akin to increasing the number of turns on a transformer and will increase magnetic field pick-up.

Effects in Gauges
Electromagnetic fields will act on the strain gauge in exactly the same way as on lead wires. The 'cures' are also similar. For electric fields, screening with, for example, aluminium foil is effective, but the foil must not come into electrical contact with the gauge. The foil could perhaps be part of the general environment protective coating. Like the lead wires, the strain gauge is an ideal sensor for magnetic fields. Screening, apart from specialised materials such as mu-metal, will not help to any great extent. A technique that has been used with success is to create a non-inductive gauge, really a continuation of the twisted lead wires. Two similar gauges are required, they are positioned and bonded at the same time so that their grids lie accurately one on top of the other. The two gauges are connected in series and then used as a single 240 or 700 ohm gauge. The principle is that currents induced in the upper gauge will be cancelled by those in the lower gauge. Such gauges must be fabricated with care, as their efficiency depends on the accuracy with which the grids can be superimposed. They can also be supplied by some gauge manufacturers.[8]

Most common gauge foils, such as constantan and nickel–chrome are susceptible to magnetic fields. Isoelastic alloy is, however, more susceptible than others and should not be used in areas where magnetic fields are likely to be encountered.

Another problem area with magnetic fields is the gauge–lead wire junction. When terminal strips are used, the strain relief loop should be kept as small as possible. The induced current is proportional to the area enclosed. Gauges which allow direct connection of the lead wire to the foil are preferred under these conditions.

Thermocouple Effect
Whilst not strictly an induced voltage, this is a voltage rather than a resistance problem. When a constantan foil gauge is connected to copper lead wires then a thermocouple is created whose output is about 4 mV/100 °C. Fortunately two equal and opposite junctions are made on a strain gauge and there will only be an output when a temperature difference exists between the two junctions. This could occur where large thermal

transients or gradients are involved, but should not be a problem in most measurements. It is, however, good practice to keep the solder connections as small and as equal as possible to avoid different thermal responses at the junctions. As mentioned earlier, a square wave a.c. bridge will not measure thermocouple voltages.

Noise Detection

It is vitally important, especially for dynamic work that every effort is made to minimise any noise present within the whole system. The first stage of this is to be aware that there is noise and ideally this should be investigated during a checkout phase prior to the test proper.

A very simple test is to switch off the bridge supply with all other equipment running. It is especially important to have ancillary devices going such as vibrators, motors etc. If any 'signal' is evident when the supply is off, then it can be classed as noise. It might not be clear where the noise is coming from, but it at least alerts the operator to the dangers. By switching on and off possible local causes, moving lead wires and perhaps trying different earthing points, then the noise source should become evident.

With multichannel installations it is advisable to include a check channel. This could be, in the case of a strain gauge system, a gauge in the vicinity of the other gauges but mounted so that it does not see mechanical strain, for example, on a plate loosely bonded to the surface. The aim is for that 'dummy' gauge to see the total environment that the active gauges see, except for the required parameter, i.e. strain. It should be wired in the same way and leads routed with the others throughout the instrumentation system. If noise is present then it should appear on all channels but it will be more evident on the check channel because of the lack of the true signal. A ratio of one check channel to ten active channels is recommended as a guide.

It is advisable to try and remove or minimise the noise source(s) in the early stages of the system, before the amplifier and recording stages. Filters can be used in some cases but quite often the noise frequencies are close to those required for the measurement and so filters cannot help. Filters will affect the response of the system and should be used with caution when pulses with fast rise times are being measured.

Short-Range Telemetry

As mentioned in the section 'Sliprings', another technique for obtaining data from rotating components is short-range telemetry. It has an

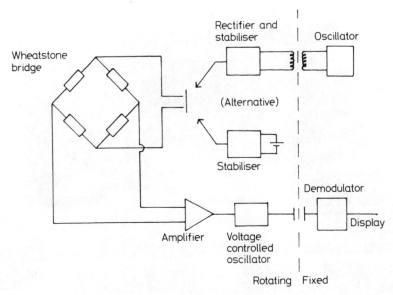

FIG. 16.   Schematic—short range telemetry system.

advantage over sliprings in being non-contacting and hence theoretically it is a lower noise system. The principles involved are illustrated in Fig. 16 and Fig. 17 shows a typical commercial system. Power for the bridge on the rotating component is derived from one of two sources, either dry cells (perhaps rechargeable) located on the rotating component, or a power transfer system. In the latter an a.c. voltage is inductively 'transmitted' from a fixed point onto the rotating part, usually over a very short distance of about 2 cm. The a.c. voltage can then be rectified, smoothed and stabilised at the correct d.c. level to feed the bridge. In either system it is advisable to use a high resistance gauge to minimise the current drain. The output signal from the bridge is amplified and used to control the frequency of an oscillator (VCO) located on the rotating part. Power for these stages will be from the same source as the strain gauge bridge. The a.c. signal whose frequency is a function of the required parameter is capacitively coupled to a fixed receiver and aerial not on the rotating part where it is converted back to the required d.c. level. The above system is suitable for single channel operation.

Systems are available which use the same principles, but sequentially scan a number of inputs on the rotating component, performing an

Fig. 17.   Commercial SRT system. Courtesy Astech Electronics Limited.

analogue–digital conversion before transmitting to the non-rotating receiver. The frequency response of the system will depend on the number of channels being scanned.

Data Links (Long-Range Telemetry)

There are instances where, for various reasons such as a hostile environment, accessibility etc., the storage of data in the vicinity of the test is difficult. A data link or long-range telemetry system is a means of transmitting a number of channels of information over considerable distances.

The radio transmitter is essentially a standard device operating on assigned frequencies, perhaps around 400 MHz. The operating distances will depend on the transmitter power output, receiver sensitivity, and the type of aerials in use at both ends. Terrain will also play an important role but the distances could be kilometres as opposed to the centimetres of short-range telemetry.

The analogue data from the transducers is usually scanned sequentially and then converted to digital and encoded into a multiplexed form that can frequency modulate the transmitter. At the receiver the signal is decoded and can be converted back to an analogue form which can, for example, be stored on magnetic tape.

There are other means available now for data transmission. These include the standard telephone lines and fibre-optics. The latter has an advantage in the greater number of data channels that can be handled.

DISPLAY

Analogue Meters

In strain measurement the analogue meter is usually a voltmeter and more commonly of the needle and scale type. This voltmeter is generally a current-sensitive device used to measure voltage by maintaining the circuit resistance constant using temperature compensation. The full scale deflection may be preset and the resolution is usually 1 unit in a 0–100 scale range. Because of their low resolution, these meters are rarely used for quantitative measurements in strain gauge testing. Their simplicity and low cost make them ideal when incorporated in instrumentation systems for signal/no-signal indication, calibration, or as null-balance indicators. Their use is most common in control rooms where analogue meters are connected to individual constant or slowly varying strain gauge outputs, and the operation process being controlled can be surveyed by a visual inspection.

In a dynamic situation, the response time of the meter and the visual response of the viewer prevents faithful reproduction or interpretation of high speed transients. A suitably damped meter will follow a dynamic signal and, depending on signal amplitude, can provide a qualitative assessment of the dynamic signal up to a frequency of 10 Hz. The damping of the needle movement can be increased preferentially for the reducing signal. A peak-hold unit results and the maximum strain/signal is displayed for a period sufficient for recording purposes, or until this signal level is exceeded.

Mechanical Digital Meters

Portable strain indicators using a manual null-balance system often have an analogue meter as the null-indicator, and a mechanical digital display in microstrain, geared to the nulling potentiometer. Resolution is one microstrain over a very wide strain range, and such instruments are ideally suited to field service conditions with their rugged construction coupled with accuracy and reliability (Fig. 11).

Electronic Digital Display

The most commonly used method of displaying a static signal is by a digital voltmeter (DVM) which comprises some sort of analogue-to-digital converter (see under 'Real-time Processing') plus some means of visual display. The display range is usually either ± 1999 (a $3\frac{1}{2}$ digit display) or $\pm 19\,999$ (a $4\frac{1}{2}$ digit display).

In strain measurement the $4\frac{1}{2}$ digit display will give the required range and resolution for general purpose testing. For transducers, a $3\frac{1}{2}$ digit display in the required units will generally be suitable. If the strain range is not known prior to testing, then manual or automatic full scale range changes are necessary. Decade range changes are usual with a zero character added mechanically to the end of the display if the operator requires the display in the same units. A range change is associated with a loss of resolution which should be avoided but can be acceptable at high strain levels. DVMs are available in a wide range of capabilities to measure signal voltages from a microvolt to ten volts to include the signal ranges experienced in strain gauge testing. DVMs capable of measuring a microvolt signal can be used without any amplification being needed.

The time required to convert the voltage into digital form and display it varies according to the methods used, to give speeds of operation between 0.5 and 10^6 operations per second.

The accuracy of a DVM is usually ± 1 unit in the least significant digit which is also the resolution of the instrument. The display, however, may only be meaningfully used in static or quasi-static testing, when only the least significant digit changes between samples. Analogue signal processing (see under 'Real-time Processing') can be employed to convert the DVM into a peak-hold unit for use in dynamic testing with the same result as for the peak-hold analogue meter. The display will only respond to either increasing, or decreasing voltages and indicate either the most positive, or most negative, signal currently measured in the test.

Numerical Printout

The coded digital signal produced in a digital voltmeter can be outputted to a printer for display purposes. Strip (2–4″ wide) printers or line (8–14″ wide) printers may be used either of the impact or thermal type. Impact printers are slower (one reading per second) because of their mechanical operation, but recording paper is cheap. Thermal printers are faster (5–10 readings per second) and quieter, but the heat sensitive paper is more expensive and the record is less permanent.

The printer is generally used for storage purposes (see under 'Storage of Data') but up to the maximum speed of 10 readings per second, the printer is a useful method of display. Its advantage over the DVM is that more information can be displayed in a more readily understood form. This extra information can include the channel number, time of reading, range indication, over-range indication, and user units. Such information allows for comparisons between readings to be made.

Oscilloscopes

The amplified signal can be fed to an oscilloscope which gives an XY display on a phosphorescent screen. The strain gauge signal is usually displayed on the Y axis (i.e. vertically) and the X axis is used to display time, or the input from another device. As the screen is quite small, typically 8×10 cm, the absolute resolution (i.e. 1 mm) is low so that oscilloscopes are used primarily for dynamic testing where their high frequency response (up to 100 MHz or more) can be utilised. However, with integral amplifiers and offset controls, a high resolution may be achieved.

The oscilloscope is most useful with repetitive or cyclic inputs which will constantly refresh the image on the screen, and which can be examined in detail over the complete cycle by adjusting the time base and phase control of the instrument.

The standard oscilloscope has a low persistence screen which means that the image fades quickly and transient events must be captured with an oscilloscope camera. However, high persistence screens and memory-scopes are now commonly available and these are much more suitable for strain measurement. With the features of analogue-to-digital conversion and digital data storage, modern oscilloscopes with integral processor can overcome the normal limitations of this technique.

Dual input channel oscilloscopes allow two inputs to be displayed simultaneously with the same time base setting. The two displays are formed by alternate switching to the two inputs.

UV Recorders

In an ultraviolet (UV) recorder (Fig. 18) the amplified signal voltage deflects a galvanometer (i.e. a moving coil meter as discussed under 'Analogue Meters') with a tiny mirror attached, which directs a beam of ultraviolet light on to moving light-sensitive paper to produce a trace of the signal variation with time. This display and recording method would normally only be used for dynamic tests, with frequencies up to 4000 Hz a possibility. The paper speed is variable and the feed continuous so that it can be speeded up during recording to capture higher frequency signals, or to examine them on an expanded time base.

The only static testing carried out with this instrument would be the static calibration for a dynamic test. The choice of galvanometer and the

FIG. 18. An ultraviolet (UV) recorder. Courtesy SE Laboratories Limited.

shunt and series resistors in the associated circuitry for a particular test depend upon:

(a) The dynamic response by which the galvanometer is rated and which must not be exceeded by the maximum test frequency.
(b) The linear range of the output trace which decreases as the frequency response increases.

Calibration of the output trace must be carried out in order to determine the linear sensitivity and frequency range. It is also possible to assess the change in sensitivity in the non-linear range. With the incorrect choice of galvanometer and circuitry, overshoot, with the trace exaggerated at its peaks, or undershoot, with the trace underestimated at its peaks, can be produced.

The UV recorder is generally a multichannel system with up to one input channel per centimetre of paper width. Recorders with 6, 12 and 20 channels are the most common. As the paper is contacted only by beams of light, different channels may produce traces at the same position on the paper at the same time. Therefore, there is no offset in the scaling of the time axis for different channels (which makes analysis simpler). The amount of trace overlap of different channels is usually kept at a minimum by the user, with a resulting lower resolution.

One or several channels may be used as manual event markers with a switch used by the operator to complete a simple battery circuit. The event marker may be automatically controlled in a repetitive dynamic test.

A grid is put on to the paper by fixed light beams within the recorder to assist in the analysis of the traces. The resolution in a typical case can be between 0·5 and 1·0 % of the full scale deflection. The time base is recorded on the paper as transverse lines at, for example, one second intervals.

Because of the exposure time required for the light-sensitive paper, the UV recorder as an on-line method of display is restricted to low frequency tests.

The latest UV recording systems do not use magnetic deflection as described above, but a solid-state method.[9] A programmable light gate array replaces the galvanometer and is digitally controlled by driver electronics to selectively pass, or block light through a precision pattern of small 'light gates' (Fig. 19). The result is an extremely high order of linearity and a trace of uniform width irrespective of rise-time although the writing speed is limited by the sensitivity of the paper.

Chart Recorders

Chart recorders may use rolls of non-sensitive paper printed with a standard grid pattern and one or more pens of different colours in contact with the paper to give continuous traces. Alternatively, the trace may be produced by a heated stylus on heat-sensitive paper, with the advantage of a smudge-free record. The deflection of the pen is proportional to the signal and the preselected paper speed provides the time base.

Dot recorders have a single rotating printing head which is controlled by a scanning unit which looks at each of up to 24 channels in turn, and prints a dot on the paper for each channel in sequence, usually using a different colour, and/or an identification number, for each channel.

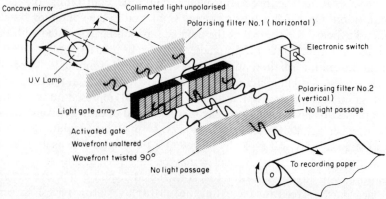

FIG. 19. A solid state optical system. Courtesy Bell and Howell Limited.

The frequency response of a pen recorder can be up to 5 Hz provided that only 10% of the chart width is utilised. Dot recorders have a very low frequency response and they are only used for quasi-static recording. Channels can, however, be commoned on one input to reduce the time interval between dots for a single input. The frequency response of a heated stylus recorder is up to 50 Hz.

When overlapping heads are used in a multichannel recorder different pens cannot mark the same point on the paper at the same time, and hence, there is a time scale offset between channels. The offset is negligible on dot recorders because of the low chart speeds.

Full scale voltage ranges are individually switch-selectable with amplification and offset usually integral in the recorder.

Visual Display Units/Graphic Screens

A visual display unit (VDU) is a low resolution screen which is a peripheral to a processor which could be controlling the data acquisition. It is usual for a keyboard to be incorporated with the VDU which enables the operator to communicate with the processor. The screen is used to display both the programmed information and the executed control commands to assist in the test operation. This information consists of alpha-numeric characters which may also be used to display the individual measured signals in strain or other units, with information relating to channel number, time etc. The characters may be used to display the real-time processed digital data in the form of a histogram or a low resolution graph. Each method of display can be constantly updated as new data is available.

A graphics translation screen (Fig. 20) is a high resolution screen which is a peripheral to the processor which could be controlling the data acquisition. Only if this type of screen is integral in the computer system will a keyboard for control of the data acquisition be associated with the screen. As for the VDU, the graphics screen can be used to display both the programmed information and the executed control commands. The real-time processed digital data may be displayed graphically in the form of a histogram, graph or even a three dimensional representation of the component behaviour. The display is constructed of dots or pixels and as the number per unit screen area reduces, so does the resolution. A low resolution display will typically represent a full angled line as a series of saw-tooth interruptions. A display may be constructed, and added to in real-time, but any modification to a constructed display can be carried out by clearing the screen and recreating the full display. With a 'refresh' screen, each dot is backed up with a unit of the computer's memory and for these

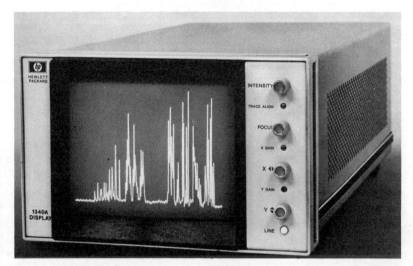

Fig. 20. A graphics translation screen. Hewlett Packard model 1340A display module in half rack cabinet, option 315. Courtesy Hewlett Packard Limited.

screens, any modification to an existing display can be carried out quickly with no need to clear the entire display. The full 'refresh' screen usually requires a large memory and increased processing power.

A VDU is only used for the indication of numeric data in static or quasi-static testing. Graphical representations may be displayed for static or low frequency dynamic testing. The limiting feature is the time taken to address the screen, and this time delay reduces the speed of data acquisition. With a dual processor system, a block of data may be displayed as subsequent blocks of data are acquired in order to indicate the manner of the signal variation. Graphic translation screens are more commonly used to display the acquired data during post processing.

REAL-TIME PROCESSING

The output of the strain gauge amplifier is a voltage which, with the correct choice of the preceding system is directly proportional to the strain change at the gauge position.

Real-time processing is carried out during the testing as opposed to post processing which is carried out remote from the test using stored data. Real-time processing ensures that the pertinent data is immediately available to the operator in readily understood units. For example, if a process has a danger level of 10 tons, the readout should be presented in tons, not millivolts or microstrain.

Analogue or digital processing may be used. Analogue processing in strain gauge testing is generally the manipulation and testing of the signal voltage—as is signal amplification and bridge balance (see under 'Wheatstone Bridge' and 'Amplifiers'). For digital processing the analogue signals are passed through an analogue-to-digital converter which provides a suitably coded output for access to a computer.

Zero Balancing and Scaling

The amplifier or the preceding switch unit, usually incorporates a variable resistor so that the initial bridge output can be balanced to zero at the start of testing. All measured voltages can thus be related directly to signal changes with no zero offsets. A gauge factor control on strain indicators (i.e. instruments with integral display units), or a gain or span control on an amplifier can be used to ensure that the signal voltage is displayed or stored in strain units. For transducer applications, a direct relationship exists between the output voltage and the measured parameter (e.g. torque, load), and the span control will provide a reading in the required units. This analogue processing has been discussed earlier.

Analogue-to-Digital Conversion

Digital data is defined as data which is in a form that is acceptable to a computer and should not be confused with the simple digital displays of commonly used strain indicators with non-electronic displays. Analogue-to-digital conversion (ADC) is carried out in specially designed circuits which are normally available as single microcircuits. A digital voltmeter (DVM) is a special application of ADC and may provide a digital output in addition to the displayed result. In static testing DVMs are commonly used to display the signal voltage. In dynamic testing, a high speed ADC is used without any display of the reading.

Analogue-to-digital conversion can be attained by the use of either integrating, or computing DVMs.

The *integrating DVM* uses a voltage-controlled oscillator to convert the signal into a periodic wave with a frequency proportional to the magnitude of the signal. This wave is then applied to an electronic counter for a fixed time interval to produce a coded 'digital' signal proportional to the average voltage over the time interval. This type of ADC will operate at speeds up to 200 readings per second. Integrating DVMs usually have an integral display unit and the capability of being interfaced with a computer or processor. Alternatively, the meter may be interfaced with a printer or punched paper tape unit if the data is to be processed remote from the test. Integrating DVMs usually produce a coded signal of $3\frac{1}{2}$ or $4\frac{1}{2}$ digits (see under 'Digital Processing') with each digit comprising a 1 byte coded signal.

The *computing DVM* compares the analogue input signal with a recurring sweep voltage. The range of this sweep voltage must be greater than the analogue signal. An instantaneous signal voltage is held and compared with the sweep voltage. A coded 'digital' counter commences at the start of the sweep cycle and stops when the sweep voltage coincides with the sampled analogue signal. The process is repeated after the counter has been reset.

High speed ADCs employing this technique will operate at speeds up to 100 000 readings per second. These are generally integral in a processor or computer frame in order to be able to store or process the data at the rated speed of the unit. The resolution and digital range of the conversion is a function of the number of steps in the sweep voltage and is increased as the instrumentation becomes more sophisticated.

With the analogue signal processed in an ADC, the signal may be displayed or printed in a convenient understandable form, stored in digital form for ease of computerised post processing, or processed by a computer or processor in the real-time of the test programme.

Analogue Processing

Real-time analogue processing can be carried out on the signal voltage. In static testing the auto-balance facility of an amplifier is a form of analogue processing, but usually such processing is carried out on dynamic signals.

Dynamic testing implies that a large quantity of data will result and a reduction in the amount of data stored, and a reduction in post processing time could be achieved with analogue processing.

Analogue real-time processing can be used whether analogue data storage, or digital data storage, is used. In the latter case, the analogue processing is carried out before the ADC and, as analogue processing is generally faster than digital processing, the total processing time can be decreased and the maximum sample rate increased.

The analogue processing can be interactive, with the response of one channel causing an operation on other channels. Typical applications of analogue processing are as follows.

Threshold Testing
A threshold voltage can be preset and the signal compared with it. The recording of data will only be initiated when the signal exceeds the preset voltage.

Signal Rise-Time
In a short duration, high speed dynamic test, such as impact testing, the data acquisition can be restricted to the event without pre-event and post-event information being acquired and causing storage problems, especially for digital data. Monitoring of the analogue signals can be carried out to test for a signal change in a short period of time. The signal change value can be preset and the time for this change can be set to a fraction of a millisecond. Upon sensing this signal change, the data acquisition can be initiated.

Integration and Differentiation
Useful analogue processing can be carried out by using integrating and differentiating circuits if energy, speed or acceleration parameters are required.

Differentiating circuits can also be used to detect turning points in the dynamic response, i.e. when the rate of change of the signal is zero. A reading of the signal can be initiated at each turning point if signal amplitude rather than frequency is required.

Signal Addition and Subtraction
Signal voltages from different gauges can be added or subtracted to produce one analogue signal. This is a useful technique to determine the signal sum and signal difference in two recording channels, and can be used to output signals directly proportional to the axial and bending forces respectively, from two strain gauges.

Voltage to Frequency Conversion
Voltage to frequency conversion can be carried out on the signal. Parallel analogue channels of signal magnitude and signal frequency with time can be obtained. This is a useful feature, especially in variable high speed dynamic testing.[10]

Digital Processing
Once the digitised signal is stored within the memory of a computer, the capability of real-time processing is limited only by the restrictions of sampling speed and the capacity of the data storage.

Both these restrictions are more pertinent to dynamic testing. In order to reproduce the dynamic response to within $\pm 3\%$, a minimum of 10 readings per cycle is required. With variable frequency testing, the sampling rate must be based on the maximum frequency. As the maximum frequency increases, the available processing time between samples decreases. Using a common processor for a number of strain gauge channels, the available processing time will decrease as the number of channels increases. Generally, only the sampled data for relatively short tests can be stored within the memory of computers other than mini and main frames. The following techniques for increasing the processing speed and reducing the required data storage can be used.

Fundamental Considerations

For the processor to be able to obtain the digitised signal or variable with the shortest possible elapsed time, five parameters must be considered:

(1) The processor's speed of operation is controlled by an in-built oscillator/timing mechanism with a fixed frequency from 1 MHz upwards. A processor must be chosen with an oscillator frequency which is adequate for the required speed of operation.

(2) If the processor's program is written in low level machine code/ assembly language, the running speed is far faster than with the high level computer languages. Maximum specified analogue-to-digital conversion rates can only be obtained with low level machine code. Of the higher level languages the compiled FORTRAN is up to 40 times faster than the BASIC interpreter.

(3) The encoded digital signal available from an analogue-to-digital converter must be acquired by the computer or processor and so must enter the random access memory (RAM) of the computer. The data can be acquired and stored within the RAM as steps in the computer program/software but the execution time of this procedure is much slower than the facility of direct memory access (DMA). If a computer has this DMA facility an area of the memory called a data buffer can be reserved and the execution of one program statement will cause the encoded signals from an input/ output port (ADC) to directly enter this memory buffer until it is filled. Data acquisition must then stop whilst this data is processed. The use of two buffers and two processors may increase the speed of operation by enabling one buffer of data to be processed whilst the other buffer is being filled. Such dual processor operations are only suitable for processing which allows for separate blocks of data being analysed separately.

(4)   The ADC circuit board can be either connected via an input/output port to the main communication bus of the processor or remote from the computer and connected via an interface cable. The transfer rate of encoded signal from the output port of the ADC into the processor's memory is increased by the first method.

(5)   The sampling speed of analogue-to-digital conversion is dependent upon the source impedance.

Function Solutions

Processing generally consists of the substitution of the data into particular equations to convert the data into the required parameters. The operating speed can be increased by special programming techniques including the maximum usage of assembly language subroutines, omission of the slower operators such as multiplication and division from the algorithms, using a stored array of data to represent a graph rather than an equation, and using standard data and program routines on read only memory (ROM) modules.

Dynamic Reproduction

To reproduce a dynamic performance to within 3%, ten samples are required per cycle. The amount of information that can be stored in a computer's memory is directly proportional to the frequency of the signal and also the number of channels of gauge signals. With 30 K byte of available memory after program requirements on a standard 64 K RAM, then 10 sec of testing at 300 Hz for 1 channel or 5 min of testing at 1 Hz for 10 channels can be stored. Although the storage is limited this is a useful method of acquiring data especially for high frequency short duration, single channel testing. A common method of obtaining useful information for a longer real-time period is to employ a threshold signal level. The signal is digitised at a constant rate of 10 times the maximum frequency and each sample is compared with the threshold level. Only if this level is exceeded will the data and a counter be stored. The counter is incremented as each set of readings is taken and so enables the time at which each stored reading was measured to be determined. Also discrete blocks of continuous data can be recognised. This technique is particularly useful in a long-term, unattended test in which there is a significant down time of the equipment or periods of low level loading which need not be recorded.

Data Reduction

The most efficient use of real-time processing is to reduce all input signals for each channel to a single variable which is continuously updated as the test continues. This variable may be the maximum strain measured during a test or the number of times that yield strain or some other such level was reached.

Turning Point Detection

If the dynamic signal is of a repetitive form but with variations in amplitude and offset then only the turning points need be stored in order to reproduce the signal. In order to understand the component behaviour more fully the complete relationship between different channels, especially the elements of a gauge rosette, is required. The use of turning points may only be made if the different signals are either in phase or 90° out of phase with all signals having turning points at the same instant of time. For such a case a detected turning point on one channel would cause all signals to be sampled at the maximum speed of the ADC. These turning points may be detected by analogue processing or by digital processing with digitised samples being made at 10 times the maximum frequency for a comparison of the successive strain changes. In a multichannel system, readings must be made at each instant from all channels with only the first channel tested because a turning point is only detected after the event. This software routine especially for multichannels makes digital processing to determine turning points only suitable for low frequency dynamic signals, i.e. less than 10–30 Hz. For higher frequencies in excess of 100 Hz analogue processing is at its most accurate and more suitable for turning point detection.

Rainflow Analysis[11]

The turning points of a dynamic response can be used in a real-time rainflow analysis in which the elements of a fixed size of array in memory (i.e. adjacent memory locations) can be successively incremented in order that the number and amplitude of complete signal cycles can be determined in post processing and used in an assessment of the fatigue damage incurred. The memory requirements for a rainflow analysis are minimal and within the capacity of many small computers. The signal change between successive turning points can be progressively converted to:

(a)   strain—using sensitivity;
(b)   stress—using non-linear mechanical properties;
(c)   critical location stress—using stress concentration factor.

The array can be one dimensional with the position in the 'column' of data proportional to the stress or strain change. However, the array could be two dimensional with 'rows of data replacing a single element in a column' of data. The location within a 'row' of data is proportional to the mean or maximum stress or strain of the signal change. The sophistication of the analysis is required when the component material has work hardening properties and the simple approach can be satisfactorily used in welded components. With increasing sophistication there is a general increase in accuracy of the resulting life prediction.

Such an analysis technique due to its fixed memory requirements and no real-time storage requirements is suitable for long-term, unattended data acquisition. The mean frequency of the dynamic signal and the length of the test will allow for an indication of the total number of cycles expected. It is important that each element of the array uses enough memory in order to store this maximum number of cycle occurrences. Three or four byte elements with maximum integer storage of 15.7×10^6 and 4.3×10^9 respectively are commonly used.

Level Crossing Analysis
Another common technique which is used at an intermediate stage in the data reduction for fatigue life assessment is 'level crossing'.[12] This technique uses a fixed memory requirement consisting of a one dimensional array with each element corresponding to a particular signal level.

The dynamic signal is digitised at a minimum speed of 10 times the maximum frequencies and the array element corresponding to the magnitude of the digitised level is incremented. If successive digitised samples are within the same signal range apportioned to that array element no incrementing is carried out. Only when the signal level changes into the signal range of the adjacent array element will further incrementing be carried out.

External Control
Real-time processing can be a form of process control. The data acquisition can be initiated by an external event or signal, and the processed data can be used to control the test or initiate an alarm system. Depending upon the equipment this control can be either digital or analogue.

Data Communication
In real-time processing the data may be multiplexed which is defined as the combination of two or more signals into a form that can be transmitted on a single transmission link and then decoded as separate signals at the receiving point.

Time division multiplex (TDM) or frequency division multiplex

(FDM)[13] may be used. The data is coded into the multiplexed form prior to transmission and can either be decoded after transmission in real-time, or stored in order to be decoded in post-processing.

In FDM (Fig. 21) each analogue signal is translated to occupy a new position in the frequency spectrum by modulating each signal on to a different frequency carrier. The signals resulting from this modulating process can then be combined for transmission. This technique may be used with measurement data in analogue form and is the usual method of telephone network audio communication.

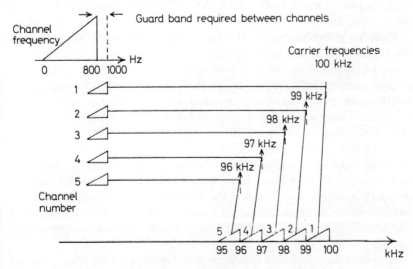

FIG. 21. Frequency division multiplexing. Translation of five separate channels each 1 KHz to occupy the frequency region 95 KHz to 100 KHz.

TDM, of which pulse code modulation (PCM) is a particular form is generally the most suitable method for the multiplexing of measurement data.[14-16]

In PCM multiplexing, the analogue signals of each channel are sequentially sampled and coded into a digital form. The digitally coded signals resulting from each sequentially sampled channel are transmitted as a serial bit stream after the addition of synchronising information.

The multiplexed signals may be transmitted along cables or by radio telemetry. If cable transmission is used repeater amplifiers are necessary at intervals to compensate for the signal loss along the cable.[17]

For transmission of data on the telephone network modulation/demodulation (MODEM) units are needed to interface into the telephone transmission system.

Gross inaccuracies, distortion and limitation of dynamic range can result from the multiplexing and transmission of data in analogue form. This results from intermodulation due to non-linearities in the translation equipment and amplifiers and system noise. With well maintained equipment a dynamic range of 40 dB may be possible (resolution 100:1).

Multiplexing of signals into a digital form overcomes the noise and distortion constraints of analogue methods. The analogue signals are presented by '0' and '1' states of a digital code which are not subject to the same noise and distortion problems.

Pulse code modulation is the more usual technique for data communication and is discussed in more detail. As all the necessary information to describe an analogue waveform may be carried in a number of discrete samples, such a digital technique is made possible. In the practical multichannel system brief samples of a few microseconds duration are gated out of each signal in turn. These are very accurately timed by a crystal controlled clocking circuit to ensure repeatability of the signals at the receiving end. This results in a train of pulses each of the amplitude of its associated sample which is a time division multiplex (TDM) signal. The analogue samples are passed through an ADC to convert this TDM signal into a PCM signal. A digital word is produced for each sample giving its value in binary notation. The word length (i.e. number of bits) determines the resolution and the dynamic range of the transmission/recording system.

Output from the digitising section is often in the form of a series of parallel words, i.e. with a separate wire for each bit. These signals are then converted into bit-serial form for transmission on a single wire, with the appropriate pulse coding to denote the logic states—'0' or '1'—according to the design of the system. With the transmitted set of data there is also a synchronising word which enables the transmitted time coding to be restored at the receiving end of the system.

At the receiving end of the PCM system the clocking pulse rate is derived from the signal itself, and the data channels within a set are identified sequentially from the synchronisation word. The digital data words are passed to a digital-to-analogue converter (DAC) to reconstitute the analogue data signals in their original form on individual outlets. One stage in the recovery of the signals involves conversion of the bit-serial PCM signal to bit-parallel form. This is suitable for direct application to a

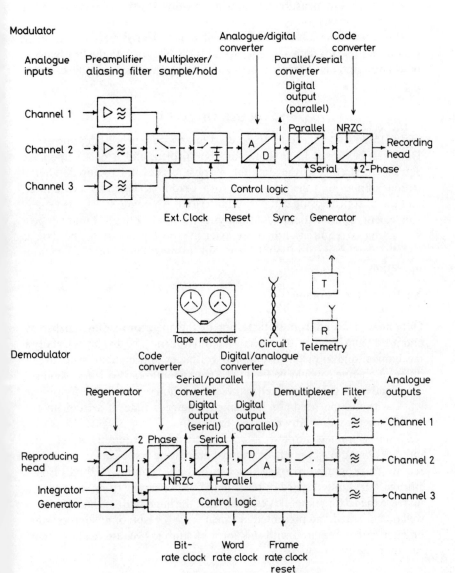

Fig. 22. Block diagram of PCM system. Time division multiplex.

computer, and in measurement data acquisition systems it is available at a computer interface outlet.

The block diagram (Fig. 22)[15] is of a practical three channel PCM system suitable for data acquisition and transmission with the link between modulator and demodulator shown as either a radio link, a cable or a tape recorder.

STORAGE OF DATA

Data stored during a test can be used in any post processing operation. Data can be stored in analogue or digital form, and depends upon the equipment used and the amount of post processing required. Digital storage implies that the data is in a form readily accepted by a computer, and the post processing can be computerised. Analogue storage is usually convenient for immediate visual analysis and if only a small amount of post processing is required. For more extensive post processing, the data is usually converted into digital form and, in most cases, this is a tedious operation.

Analogue Data Storage
Manual
Data may be stored from static testing by the operator reading the display and producing hand written sheets in tabular form. The display is usually a mechanical digital meter or a digital voltmeter and manual switching of different channels would be used. The equipment required is less sophisticated and more portable which makes this method more suitable for on-site work without the need for instrument engineer time in setting up the system, or providing any required back-up.

If only a limited amount of post processing is required, or the total number of readings is small, or the post processing is subjective and cannot readily be programmed, then this can be an acceptable method of data storage. The limited speed of reading the display and writing it down will increase the testing time. It is essential that the hand written sheets are very well documented and presented in a neat form for ease of post processing. Greater care is required with this form of storage because reading errors can more easily occur.

UV Recorder
The UV recorder has been described previously as a means of displaying the output of dynamic testing. The trace can be used as a permanent record of the test and for post processing purposes. However, as there is a gradual

degradation of the trace caused by continued exposure to daylight, for permanence the trace should be chemically fixed.

For high frequency, short duration tests, this form of storage is usually used as a permanent record for back-up purposes to a 'blind' method of storage. A 'blind' method of storage would be digital or analogue magnetic tape recording in which there is no visual appreciation that data in the required form is being recorded. As a stand-alone method of storage, the disadvantages of the UV recorder include low resolution, variation of trace intensity and difficulty in recognition of cross-over traces. The reduction of the trace into a form suitable for post processing is usually manual, although short traces can be digitised on a computer peripheral. By both methods, the acquisition of any detailed information is tedious.

For low frequency, long duration tests, this form of storage is a useful method. A major consideration is the paper speed. For signal frequencies from static to 0·4 Hz the roll of paper will be sufficient for 24 h operation before a replacement is necessary. At this paper speed, the dynamic signal can be fully interpreted and the relationship between the response of different channels appreciated. As the frequency of the signal increases, either the paper is replaced more regularly, or the trace becomes a band with only the upper and lower limits definable. In some tests, the data can be recorded for timed intervals in order to reduce the amount of paper consumed and minimise attendance at the test. The reduction of the trace into a form suitable for post processing must be manual. It is very tedious, and requires good on-site documentation.

Chart Recorder

The chart recorder has been described previously as a means of displaying the output of a quasi-static or low frequency test. The trace can be used as a permanent record, or for post processing purposes. The reduction of the trace into a form suitable for post processing is generally manual. In comparison with UV recorder traces, the reduction is less tedious due to the larger paper width and the different colour of ink for different traces. However, the time scale offset between traces does cause inconveniences.

Chart recorders are usually used for long-term unattended tests with a small number of channels. As the number of channels increase, the multichannel capacity of UV recorders becomes more economical.

Numerical Printout

The numerical printout from a static, quasi-static or low frequency (< 1 Hz) dynamic test can be used as a permanent record of the test and for post processing purposes. It is common for a numerical printout to be obtained as a display and for security purposes if a 'blind' digital storage method is employed.

As a stand-alone method of data storage, it is comparable to hand written sheets with an associated increase in accuracy and the disadvantage that the data cannot be directly produced in tabular form for ease of understanding.

Photography

The display unit can be photographed to obtain a permanent record of the test which can be used in post processing if the data cannot be stored in a suitable manner due to the limitations of the existing instrumentation. Photography is the standard method of storage with oscilloscopes and attachments for single frame or high speed cameras are available. The method of storage of data displayed by oscilloscopes is the main disadvantage of this class of instrument. To overcome this problem, three types of oscilloscopes are now available:

(a)   Oscilloscopes with integral light beam recorders in which the screen display is recorded on the light-sensitive paper.
(b)   Oscilloscopes with storage tubes in order to retain the trace on the screen.
(c)   Oscilloscopes with integral ADC/DAC processor and digital data storage such as the Nicolet Explorer.

Another case in which photography can be used, is the recording of low frequency dynamic or quasi-static data, with DVM displays. If the test equipment is several channels of single channel instrumentation designed to give a display only, and the output is slowly changing, then the instantaneous readings for all channels cannot normally be obtained. Provided that the display is automatic in that no manual balancing is required, then the DVMs can be stacked and photographed during the test. The least significant digit of the DVM display which may not have been readable by eye due to its rate of change can be obtained and the instantaneous relationship between the outputs determined.

Analogue Tape Recording

The analogue data can be recorded on magnetic tape by using a reel-to-reel recorder, or a cassette recorder. The full scale signal voltage range for each strain gauge channel can be preselected and the signal recorded with each channel using a different track on the magnetic tape. Within the operating signal frequency range of the instrument, the accuracy of the reproduction of the signal is good and at the recording stage, provided that the environmental conditions are good, the resolution can be better than 1 % of full scale.

The recording head is similar to a transformer with a single winding in which the signal current flows to produce a magnetic flux in the core material. The core is made in the form of a closed ring with a short non-magnetic gap in it which is bridged by the magnetic tape to complete the magnetic path. The strength of the magnetic field is directly proportional to the rate of change of the signal current.

Direct recording of the signal voltage can only be made if the signal is constantly changing. If the rate of change of the signal voltage is zero, i.e. a static signal, the strength of the magnetic field at the recording head is zero. The frequency range for most instrumentation tape recorders with direct recording, is usually between 50 Hz and 600 KHz (intermediate band).

Frequency modulated (*FM*) recording may be carried out on the modulated signal voltage. Frequency modulation is accomplished by varying the carrier frequency in response to the amplitude of the data signal. The recorded voltage will be constantly changing even for static testing. The frequency range with FM recording is usually d.c. to 40 KHz (intermediate band) and d.c. to 80 KHz (wide band).

A problem with FM recording is the increase in signal-to-noise ratio and hence the decrease in resolution.

Reel-to-reel recorders (Fig. 23) can have 4, 7, 8, 14, 32, 42 or 64 tracks. For each channel, there is the facility of preselected input voltage range and the option of direct or FM recording. The accuracy is optimised by reducing the effect of errors such as the non-linearity of the variation of magnetic field with input signal voltage, hysteresis losses in the core material of the recording head and the variation in the magnetic field towards the edges of the non-magnetic gap. A range of tape speeds in the ratio 128:1 is available with either 15/16 to 120 inch per second or 15/32 to 60 inch per second.

It is possible to take data from a 'reproduce head' upstream of the 'record head' to allow for examination of the data recorded on the tape by a display unit such as UV recorder or oscilloscope.

Less sophisticated post processing equipment is necessary as high frequency dynamic recordings can be replayed at slower speeds. In long-term testing the recording can be at the slowest tape speed for the signal frequency, in order to keep tape usage and test attendance to a minimum, but post processing can be carried out using a faster tape speed.

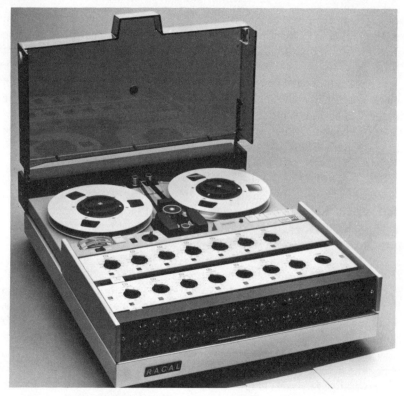

FIG. 23. A 14 channel instrumentation tape recorder. Courtesy RACAL.

Cassette recorders usually have 2 or 4 tracks with a common fixed signal voltage range. The cassette recorder will utilise either direct or FM recording common to all channels. The type of recording is usually fixed in cassette recorders. 'Record' and 'playback' speeds can only be varied with a limited range in good quality cassette recorders. Usually the same tape speed is used for 'record' and 'playback' in the same cassette unit. With cassette recorders, it is common for a different playback unit to be necessary if a different tape speed is required. The dynamic range of cassette recorders is restricted. Cassette recorders are smaller and more portable than reel-to-reel recorders, and also much less expensive. If the reduction in features of the cassette recorder can be accommodated as in long-term, unattended, low frequency testing for a small number of channels, then the cassette recorder will be a suitable and accurate method of data storage.

When data is being recorded on to magnetic tape, it is recommended that whenever possible, an analogue meter or UV recorder is used in order to see that 'reasonable' data is being recorded. The UV trace could also be used as a security back-up.

The preceding discussion refers specifically to *instrumentation tape recorders* in which sources of error such as (a) non-linearity—due to tape medium, record/reproduce head transfer characteristics and non-linearity within the electronic circuitry; (b) noise—due to the recording system electronics, imperfections of the tape and tape movement system; (c) drift—due to the instability of the electronics with temperature and power source changes; and (d) time correlation—due to interchannel displacement errors are minimised within the specified signal and dynamic ranges.

Audio tape recorders can only be used to record analogue data signals with any degree of accuracy for particular frequency ranges corresponding to the mid-audio range, and care must be taken in the choice of their application.

Digital Data Storage
Punched Paper Tape
Data can be stored on punched paper tape with the eight bits of each character (i.e. one digit is a character) represented by a hole or no hole corresponding to a value of 1 or 0 across the width of the tape. The representation of the characters with particular binary configurations usually corresponds to the ASCII code. Punch units usually operate with a standard serial or parallel interface. Punched tape is a convenient form of data storage if no computer or processor is included in the instrumentation system. The punch unit may require a conversion interface from the BCD output of the DVM. A system utilising punched tape would also include a printer which is used for display purposes during the test, and the printout could be retained as a security back-up. If a computer is integral in the instrumentation, more convenient forms of digital data storage are available.

The maximum speed at which the paper may be punched is usually 75/110 characters per second which, depending on the system, is equivalent to 4–6 readings per second. This restricts its use in a multichannel system to static testing. The punched tape can be read into a computer for post processing. With desk top or microcomputers, the reading of punched tape can be carried out with conventional peripherals at speeds up to 120 characters per second, or 8–12 readings per second. The reading speed can be increased by a factor of 10 with the paper tape reader peripherals of main frame computers.

Difficulties can occur in the reading of punched tape because of punching errors due to sticking paper feed, deviations from the expected order of the data, and reading errors due to sticking paper feed. These difficulties make the reading of punched paper tape a tedious operation.

Certain punch units are very robust for site conditions, and simple to operate, with features such as a take-up reel for used paper, automatic winding-on of new paper rolls and buffered data to compensate for short-term differences between reading and punching speeds.

Punched tape is becoming less popular as processors and computers (with their more convenient magnetic digital storage) are used more widely in strain gauge instrumentation systems.

Solid State RAM Storage

If the instrumentation system includes a computer or processor then the digitised signal is read either through an interface from a DVM or direct by an ADC and must enter the random access memory (RAM) of the computer. Depending upon the speed at which readings are required and on the amount of real-time processing, the digitised signals can either be stored in the RAM until the end of all, or a discrete part of, the testing, or they can be transferred immediately to display and/or storage peripherals. In the latter case there will be no restrictions produced by the size of the RAM.

The memory size is stated in either bytes or words. One byte corresponds to eight binary bits of data and a word can consist of between one and twelve bytes. The RAM can vary from as little as several hundred bytes in certain microprocessors, to several million bytes in main frame computers. With a resolution of unity, the reading range of one and two byte digitised signals are ± 128 and $\pm 32\,768$ respectively. This range is obtained by raising 2 to a power equal to one less than the number of bits employed.

One byte readings are sufficient for most dynamic testing whereas two byte readings are sufficient for most static testing, thereby giving comparable resolution with analogue forms of recording. In desk top computers and microcomputers in which no assembler facility is available, the reading must be stored in whole bytes, whereas, if the assembler language facility exists, then the number of bits per reading can be chosen.

RAM data storage is not permanent and requires constant power in order that the data is maintained. For different types of computers and processors this problem can be overcome in different ways.

Desk top computers or microcomputers can have an inverter in the power supply to use a battery back-up in the event of power failure. The status of the battery can be checked by timing its use within the computer program and after a time setting, reducing the computer's power requirements by stopping the acquisition and processing and hence, only maintaining the RAM.

Alternatively, the program and the initialising data such as datum readings and gauge factors, can be stored on magnetic disc or tape if the computer has an auto-start feature upon return of power. The auto-start feature loads in usually the first programme on the disc or tape, and operation of the programme is initiated. This method of maintaining data is only possible if the initialising data remains constant and is not a function of the readings, and if each set of data is transferred from RAM to permanent storage between sets of readings. A power failure will then result in only the current set of readings being lost.

A microprocessor based system which is purpose-built for strain gauge testing, could incorporate dry battery memory back-up on the RAM printed circuit board. The type of RAM chip used will govern the maximum memory size, the speed at which data can be read to or written from memory, and the power requirements to maintain data in the RAM and to transfer data to and from the RAM. With current chip technology, a small battery could typically maintain data in a 24 K byte RAM for one month. As the power requirement reduces to very low levels, however, the speed at which data may be written to or read from the RAM reduces significantly.

In a purpose-built microprocessor system, the RAM board, with integral battery, could be in a removable sealed unit to facilitate data acquisition on-site, and replacement when the RAM is filled, and for post processing on a centralised computer.

Digital Tape Recording

Storage may be on a cassette or cartridge which is either integral in a desk top/microcomputer or in its peripheral device. Alternatively, it could be a reel-to-reel recorder which could be a peripheral to a main frame computer or processor.

Reel-to-reel recording may be carried out using either a digital tape transport or an instrumentation recorder. *Digital tape transports* are computer compatible peripherals to minis or main frames with 7 or 9 tracks, 1600 bits per inch and a 45 in/sec tape speed. This type of recorder is designed for stop/start operation as data is recorded record by record. Usually the multitask capability and large memory of main frame computers enables the test data to be stored in the RAM and digital tape transport peripherals would only be used to post store the data. *Instrumentation tape recorders* as used for analogue data may also be used for digital data as a peripheral to a processor or even a main frame if the data acquisition rate and the amount of data exceeds the memory capacity and the slow transfer speed to digital tape transports. These recorders operate continuously with up to 42 tracks and 24 000 bits/in/track. Tape speed is variable up to 240 in/sec and the standard $10\frac{1}{2}$ in diameter reel is of length 4600 ft. The sources of error in analogue tape recording are eliminated when suitably coded digital signals are recorded.

As with analogue storage, it is also possible to examine the recorded data at a 'reproduce head' upstream of the 'record head'. Digital-to-analogue conversion is required before displaying the data on devices such as a UV recorder or oscilloscope.

A cassette or cartridge is a method of data and programme storage having the tape drive unit integral in many desk top/microcomputers. *Cassettes* are of the audio type with no restrictions as for analogue data recording. They have a storage capacity of the order of 60 K bytes and are relatively inexpensive. *Cartridges* are used for high density data storage with a capacity of 250 K bytes and high speed tape movement. The cassette or cartridge drive unit could be a peripheral to a small computer or processor. Larger cartridges could be used in a peripheral with storage capacity up to 1 M byte. Multidrive tape units can be used to give increased flexibility and greater storage capacity.

On the tape, the data is stored in records which are usually 128 or 256 bytes and a number of such records constitute a data file. Data is transferred to the tape either via an IEEE interface or directly at speeds usually of 1200 bytes per second but speeds approaching 10 000 bytes per second can be achieved with certain units. The total time required to open and close a file is about one second. The time taken to find the required file depends upon whether the storage is sequential or random. Sequential file storage produces data stored on the tape in the same order as it was transferred. With no file access time this is the fastest method of storing data on tape and is especially useful when storing between sets of readings for a single processor instrumentation system. A further increase in speed is achieved by buffering data either in the computer's RAM with an integral tape, or in the computer's RAM and in the peripheral with a peripheral tape unit. Random storage is the transfer of data to different positions or files on a tape in order to facilitate post processing. Due to the increase in tape movement/file access time, random storage which is produced by on-line processing of the data, is generally only carried out in static or quasi-static testing. A common use of random storage is the creating and updating of a file for each gauge in a multichannel system. In post processing a graph of strain against time for a particular gauge can be obtained after loading the relevant file. The alternative would be to search through large amounts of data to collect that data pertinent to the particular gauge. If the tape unit is dual drive, or if sufficient space is available on a single tape, then random on-line storage need not be carried out because it can be achieved in post processing with unattended operation. Random storage requires larger amounts of tape movement with the increase in time taken to store and also the greater risk of physical damage to the tape. For these reasons, sequential storage on tape is the more common method of data storage.

Disc storage

Flexible and hard disc magnetic media are accepted methods of data storage with the main advantage being fast file access time.

Soft, flexible or floppy discs are generally 8 in diameter and made of thin plastic with an oxide coating. Smaller discs which are 5 in diameter are referred to as diskettes. The discs are an alternative to a long length of oxide coated plastic film as in cassettes. Data is recorded on, or read from, the disc by a non-contacting head which is able to move across the constantly rotating disc.

Disc drives are usually dual units to give increased storage capacity and the flexibility in operation of two discs. Drive units can have one or two heads per disc to enable one or both sides of the disc to be used. One side of a disc has a capacity of 250 K or 500 K bytes and for diskettes, this storage capacity is reduced by half.

Soft disc storage is comparable in capacity and cost with cartridge data storage and hence disc drive units are integral in many microcomputers for data and programme storage. They are also standard peripherals of other microcomputers and most desk top computers with an IEEE, or 16 bit parallel interface. The data transfer speeds and file opening/closing times are comparable with tape cartridges but the file access speed is greatly increased due to the ability of the head to scan the surface of the rotating disc quickly. Sequential data storage is no faster with discs, but random data storage is made appreciably faster so that it is more likely to be used. An important advantage with discs is that no mechanical damage is caused to the disc during file access as is the case with magnetic tape.

Soft discs are very susceptible to damage in handling and dust is a particular hazard. Therefore, care must be taken in their storage and use.

Disc drives are not as transportable as cassette or cartridge drive units, due to the recording head easily becoming misaligned.

Hard discs are oxide coated rigid discs to ensure that at the high speed of rotation—approaching the speed of sound at the periphery—the disc does not change shape and hence, the gap dimension between head and disc is maintained. The file access, file opening/closing, and data transfer times, are all reduced due to the increased speed of rotation. File access time is reduced further with multihead capability in which each track of a disc has a separate recording head. Multidisc drives can handle between 10 and 20 hard discs with a total storage capacity of several hundred million bytes. Such units could have non-replaceable hard discs and are comparable with reel-to-reel tape recorders as peripherals to minicomputers or main frame computers.

Smaller hard disc units are becoming popular as a peripheral to microcomputers. The storage capacity is approximately 20 M bytes and the hard disc is in a replaceable sealed unit. Hard disc units are much less transportable than soft disc units. The smaller units are portable, but restricted by their weight. Hard discs are more susceptible to dust and handling than soft discs hence the reason for certain units using non-replaceable discs to eliminate handling and requiring air conditioned atmospheres to reduce the risk of dust.

CHOICE OF SYSTEM

Twelve factors which affect the choice of an instrumentation system are briefly discussed in this section and several typical systems are described with reference to these considerations. When twelve factors are considered with each capable of much finer subdivision, an infinite number of possible system types can be imagined.

Test Conditions
Signal Frequency
The test frequency expressed as cycles per second (Hz) is usually expressed as the maximum frequency of signal variation which may occur for only a part of a cycle. Static test conditions (0 Hz) usually only occur when artificially created as in the incremental loading of a component.

Quasi-static test conditions assume that the strain levels in a structure under test remain constant, within the resolution of the reading whilst readings are taken from all the gauges. With a fast reading speed and few channels, a signal frequency of 3 Hz could be regarded as quasi-static. A quasi-static test usually requires the relationship between different gauge outputs at the same structure loading to be determined. Quasi-static conditions for single gauge testing are rarely required. Dynamic test conditions assume that the strain levels in the structure do not remain constant whilst readings are taken from all the gauges. A separate acquisition system is usually required for each gauge output.

Test Duration

The duration of a test should be expressed in terms of the amount of data acquired but if a test exceeds 12 h then generally the equipment must be capable of operating either unattended or with non-specialised personnel. Ease of operation is essential.

Long-term monitoring must rely on real-time processing to reduce the amount of analogue or digital data acquired. Renewal of the storage media, i.e. UV paper or magnetic disc, is therefore kept to a minimum and so too is the post processing and the physical storage of the storage media.

Number of Gauges

For a multigauge installation there may be a separate complete instrumentation channel for each gauge or certain parts of the instrumentation may be commoned to allow scanning or multiplexing. If separate channels are used, any real-time processing can assume no interrelation of the different gauge outputs.

In static or quasi-static testing it is usual for a common supply to be switched to gauges in sequence and for the signals to be switched through a strain indicator which includes signal conditioning, amplification, analogue processing and display and/or analogue storage. Whether a portable strain indicator is used with manually operated switching between channels, or a computer controlled multichannel data logger is used is mainly a question of speed and the amount of processing required during the test. For static or quasi-static testing the signal does not generally need to be read quickly but the economics of time and staff involved is the important factor.

In dynamic testing it is usual for each channel to be separate for signal conditioning, amplification, analogue processing and analogue storage. Digital processing and storage can be separate in a microprocessor controlled system.

It is also possible to use a common processor to carry out analogue-to-digital conversion on each channel in turn and to multiplex the data on a single channel of a digital tape recorder. For all systems it is possible for the analogue and digital processing of separate systems to have interrelated activity, i.e. the result of the processing of one channel can be made to affect the method of processing another channel.

Real-Time Processing

Real-time processing is carried out as the testing proceeds. The minimum processing is usually the conversion of voltage to strain, stress, or load units to enable the test data to be more easily understood by the test engineer. It is also carried out to reduce the amount of test data which makes the test easier to operate, but the result can be to facilitate or eliminate the post processing.

Attended or Unattended Test

As all testing is important, a test will be attended, wherever possible, to ensure that the instrumentation is operating satisfactorily and to observe the response of the component under test.

In certain cases, once the instrumentation is working satisfactorily, a test may be left unattended usually because the test is either of long duration or in a hostile or remote environment. The system used in these circumstances either may only require the minimum of attendance as the storage media is renewed or can be controlled by a centrally based computer via telemetry or MODEM with regular transfer of data from the system to the computer.

Accuracy/Resolution

Using the basic principles of strain gauge instrumentation it is important that readings are made as accurately as possible. The possible sources of error in a strain gauge test must be reduced as is practicable (Chapter 5). For static testing, instruments are available with a resolution of one microstrain which is as high a resolution as is usually required. No consideration is usually required of the expected strain range of the test. In dynamic testing it is essential that the expected strain range is known or estimated. With UV recorders, chart recorders and real-time analogue-to-digital conversion the resolution can be between 0·5 % and 5 % of the total strain range. The higher resolution would be obtained by analogue-to-digital conversion because there are no problems of overlapping traces. A chart recorder usually has a better resolution than a UV recorder because of the possibility of more easily changing strain ranges during a test.

Analogue tape storage will be replayed during post processing with offset and amplification possible to optimise the resolution of the analysis.

Non-Repeatable Tests
If, because of damage inflicted on the test component, or the cost/time involved, the test is not repeatable then all the required information must obviously be acquired in the single test.

Providing that the choice of system is correct static testing causes few problems as time is available to ensure that the correct information is acquired.

In dynamic testing most eventualities from loss of power supply to equipment faults must be covered by back-up systems and separate channels to minimise potential data loss.

Repetitive Testing
In repetitive testing the same test is carried out on many components, or on the same component with different parameters.

Such testing is usually carried out in order to compare the results of all the tests and it is important that the same system is used in all tests. Speed of carrying out the test and the ease of operation of the system are essential features.

If the testing is a continuing routine, for example as the quality control of a production item, custom-built instrumentation systems are often used in order to simplify the use of the system with the minimum information displayed and stored in a form readily understandable by the technician controlling the test.

Environment
The environment in which the system must work satisfactorily is very important and could be roughly divided into laboratory and on-site conditions. In the laboratory the conditions may be controlled whereas on-site the equipment must be capable of operating in a range of different conditions.

Choice of on-site equipment is made with reference to the equipment operating with different power supplies, exposed to the normal weather conditions and being portable and transportable. More hostile environmental conditions would include extreme temperature range, dust, vibration, magnetic fields, power supply fluctuations and noise. Certain hazardous conditions require that the equipment is intrinsically safe.

Security
If data storage is required whether for post processing or for record purposes it is important that, if possible, the data is acquired in two forms. A cross check can then be carried out to investigate unusual readings.

Post Processing
In many tests analogue or digital data is displayed during the tests and the post processing includes visual inspection or simple calculations.

If the post processing can be defined and if the quantity of data is sufficient then a computer can be used. If the system does not produce data in a digital form compatible with a computer then the data must be manually converted into a suitable form.

The most common post processing would be datum subtraction, gauge factor correction, strain to stress conversion and presentation of results in tables and graphs. In certain systems for short duration high frequency testing, a post processing facility exists on-site to enable test results to be obtained between tests.

Availability of Equipment
The test requirements are usually tempered by the equipment which is available. This is usually the incentive for systems to be no more sophisticated than required. Unfortunately it is also the reason why the required and necessary information is sometimes not obtained.

Typical Systems
Static Testing
A static test usually consists of loading a component or structure in increments with strains being recorded for each increment. Typical examples are loading an aircraft wing in a test rig, or proof testing a pressure vessel.

The choice of instrumentation depends on the number of channels, the available time or manpower, the time over which each load increment can be held constant for all the channels to be scanned, and the amount of real-time processing required.

Testing may be carried out under laboratory, workshop, or site conditions.

The simplest equipment would be a manually operated portable strain indicator (Fig. 11) with as many manually operated switch-and-balance units as necessary. The equipment is very portable and robust and ideal for on-site work. There are no limitations on the number of channels which can be used with such a system except those of time and manpower available. One operator can log up to 10 readings per minute on hand written sheets. Real-time processing is minimal and post processing consists of either the manual preparation of graphs, or the reduction of data using a calculator or computer.

The speed of data acquisition can be increased by using a simple strain gauge logging system which has a DVM display, automatic scanning and print out for up to 100 channels for a typical system (Vishay 220). The scanning and printing speed is one channel per second. This system may also be interfaced with a tape punch to simplify post processing by computer or interfaced with micro or desk top computers to give real-time processing and storage. The logging system is still portable, but not so robust as the manual system. Note that general purpose loggers require special input conditioning for use with strain gauges and often do not have the stability and resolution necessary for handling the low level inputs from strain gauges.

More elaborate instrumentation systems have much higher scanning speeds and sophisticated computer control of test, real-time processing, and storage. The system shown in Fig. 24(a) utilises a desk top computer which is only dedicated to the system whilst actually scanning and recording test data. At other times it can be used elsewhere for other purposes. The scanning speed of this system is 25 channels per second with a maximum of 1000 channels. Post processing and test programming can be done with the desk top computer off-site under office or laboratory conditions.

Larger centralised computers can be used for post processing either using data recorded on-site, or with direct inputs from a remote data acquisition system using a data link, such as a MODEM operating on GPO telephone lines Fig. 24(b).

Dynamic Testing

Unlike static testing, the requirements of dynamic testing instrumentation systems are varied and no common factor can be used in the discussion.

General purpose equipment. This can be divided into two groups as follows.

(1) Analogue

Such equipment would be used generally for up to 30 channels, attended testing with a frequency range up to 20 000 Hz to include strain measurement and noise vibration studies in a workshop or laboratory environment.

Post processing would include the visual examination of traces to obtain an appreciation of the component's performance before any quantified analysis is carried out.

A typical system includes multichannel amplification (Fig. 15), a UV recorder (Fig. 19) for selected channels and an analogue tape

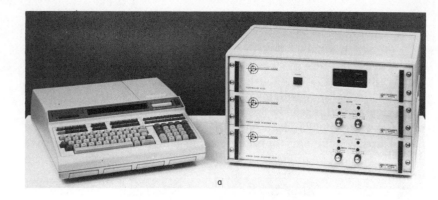

FIG. 24. Advanced computerised data acquisition systems. (a) Courtesy Measurements Group, Vishay. (b) Courtesy Intercole Systems Limited.

recorder (Fig. 25). The UV recorder can be used up to 4500 Hz but is usually used up to 200 Hz for strain measurement. It will display selected channels to give an on-site assessment of the test data. The UV recorder would be replaced by a single or dual beam oscilloscope for frequencies up to 20 000 Hz, but the most common requirements for noise and vibration measurements are in the range 500 to 10 000 Hz. The instrumentation tape recorder provides the facility for different recording voltage ranges on different channels. The recording range can be selected to exceed the expected range and if, in the post processing, amplification, offset and filtering can be carried out a resolution of $\frac{1}{2}\%$ can be achieved. The post processing can be carried out using either UV recorder or ADC and computer.

The signal conditioning/amplification include shunt calibration resistances, and a voice channel on the tape recorder can be used to annotate calibration and testing. A disadvantage of such systems is the cost of using separate amplifiers/conditioners for each channel of processing and storage.

An alternative would be to use a limited number of channels of instrumentation but the disadvantages are the time taken in repeating the testing and post processing with successive sets of channels.

The main advantage is that the analogue tape can be replayed at different speeds to enable it to be post processed by computer after an analogue-to-digital conversion, or selected sections of the test data can be visually inspected in detail at a later date. The data can also be recorded wideband, and filtered on replay to select the significant frequency bands for analysis.

(2) Digital

Large numbers of channels can be most efficiently handled by digital equipment either for recording on magnetic tape, or for transmission by PCM technique. Because the analogue signals are converted into digital form they can be handled without deterioration due to noise and distortions caused by the electronics and recording medium. The resolution of a PCM system is defined by the number of bits in the data word with 10/12 bit words commonly used, corresponding to resolutions of 1:1024/1:4096 and digital ranges of $\pm 512/\pm 2048$ respectively. Equipment suitable for attended or unattended testing in the workshop, laboratory or on-site environments is available. Bandwidth (frequency range) available per channel will be determined by the transmission method or the data density which can be accommodated on magnetic tape. PCM systems are most suitable for the acquisition, storage and transmission of signals in the frequency range from d.c. up to 10 KHz.

A typical system (Figs 25(a) and (b)) includes 8 channels of amplification, a PCM modulator in which the amplifier outputs are sequentially sampled, digitised and formed into a serial digital bit stream with the addition of synchronisation words. These systems can either be used separately to acquire 8 channel blocks of data for recording or transmission or can form part of systems providing up to 256 input channels.

The PCM demodulator takes the PCM signals from the transmission link or recorder and decodes it into separate channels, restoring the analogue form for UV or chart recorder and a digital output form suitable for direct access to a computer for processing.

Compatibility of different manufacturers requires a standard for the format of the digital bit stream and synchronising word. Care must be used in constructing a system from different manufacturer's equipment.

A range of compatible PCM equipment together with suitable transmission equipment is available to facilitate its use in difficult and remote environments such as rotating structures and test vehicles during crash testing.

Special purpose equipment. There are some instrumentation requirements which are not met by standard off-the-shelf commercial systems. However, a solution can usually be found by making up a suitable 'package' of individual pieces of equipment from different manufacturers and interfacing them. This requires expertise and wide experience, and for those who do not have this, there are specialist companies who will undertake it. Using the customer's specification for the test programme, they will assemble a package, complete with necessary interfaces, and commission the complete system. Two examples of such systems are given below.

(1) Analogue

A specific requirement is discussed because of the large possible range of equipment in this field. The requirement was for a few channels and low frequency with post processing needed. For more channels at higher frequencies, equipment using telemetry and multiplexing would be used. The system had to provide four channels with a frequency range 0–8 Hz for unattended testing with 12 h of testing between storage media changes. A resolution of $\frac{1}{2}\%$ of maximum strain range in post processing was required and the system had to be suitable for use in a hostile environment.

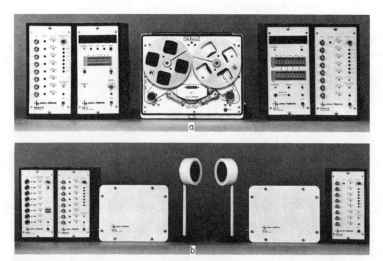

FIG. 25. A PCM (digital) instrumentation system for dynamic recording (a) with intermediate tape storage and (b) with FM/FM data link. Courtesy Johne-Reilhofer.

Four separate channels of signal conditioning/amplification and a four track cassette recorder are used on-site. A laboratory based playback unit incorporating amplification, offset and filtering is suitable for reproducing the signal on chart recorder or by a computer.

The on-site equipment is small and easily housed in sealed containers so that the unit could be contained in dry, normal temperature range conditions even in a hostile environment.

Such a system has uses in the long-term testing of wind and wave loaded structures which have typical frequencies of $\frac{1}{2}$–2 Hz. A disadvantage is the necessity for changing the cassette after 12 h of testing. If the data is only to be sampled for 15 min periods of constant loading, e.g. different weather conditions, then the recording could be initiated by a telemetry link and the cassette need only be changed every week.

(2) Digital

A specific requirement is discussed for a single channel system with real-time processing to reduce the data from all testing to an array of memory of a fixed size with each element incremented as and when the signal variation dictates. For multichannel operation a system similar to that described under 'General Purpose Equipment' could be used with multiplexing and telemetry.

The system (Fig. 26) is a solid state, microprocessor controlled, single channel instrument for short- or long-term, unattended, on-site processing and recording of analogue data in a continuously reduced format. The sealed unit contains signal conditioning, amplification and analogue-to-digital conversion. Collection, processing and storage are controlled by the preprogrammed instructions contained in interchangeable, solid state,

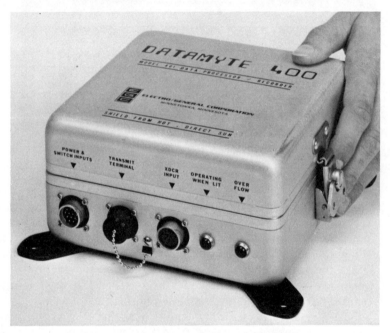

FIG. 26.   A single channel system for long-term, unattended recording. Courtesy Electro General Corporation.

plug-in units. The storage can be interrogated at any point during the recording process as it can communicate with a data terminal either directly or via a MODEM. Alternatively, the interchangeable plug-in memory unit can be removed and its integral battery will maintain the memory for up to 90 days.

The unit is compact and portable and is sealed, resistant to vibration, temperature and humidity. It may be powered by a.c. mains or 12 V d.c.

For frequencies in the range 0–200 Hz, one of the program packs will enable a rainflow analysis to be carried out to provide an array of data relating to number of cycles of each strain range which is useful for a life assessment fatigue analysis.

A disadvantage is the necessity for knowing all the detailed post processing requirements beforehand.

REFERENCES

1. PERRY, C. C. and LISSNER, H. R., *The Strain Gage Primer*, 2nd edn., McGraw-Hill, New York, 1962.
2. DALLY, J. W. and RILEY, W. F., *Experimental Stress Analysis*, McGraw-Hill, New York, 1978.
3. *Optimising Strain Gauge Excitation Levels*, 1977, TN-127-3, Micro Measurements Group, North Carolina, USA.
4. *Errors Due to Wheatstone Bridge Non-Linearity*, 1977, TN-139-2, Micro Measurements Group, North Carolina, USA.
5. DREW, D. A., Sliprings, a review, *Strain*, October 1966.
6. GALL, M. H. W., Strain gauge slip ring circuits, *Strain*, January 1979.
7. KLIPEC, BRUCE A., How to Avoid Noise Pick-Up on Wire and Cable, *Instruments and Control Systems*, December 1977.
8. *Non-Inductive Strain Gauge (Type 125 WJ)*, *Micro Measurements*, Micro Measurements Group, North Carolina, USA.
9. Solid state method for recording and analysis transducer technology, *Transducer Technology*, 3(1), October 1980.
10. DOEBELIN, E. O., *Measurement Systems—Application and Design*, McGraw-Hill, Kosabusha, 1975.
11. *A Cycle Counting Algorithm for Fatigue Damage Analysis*, SAE No. 740278, Society of Automotive Engineers, New York, 1974.
12. *Field Service History Analysis for Ground Vehicles*, Paper 750553, SAE Earthmoving Ind. Conf., Peoria, Illinois, April 1975.
13. *Multiplex Technology*, Lenkurt Demodulator, 2nd edn, 1966.
14. *Binary Logic and PCM*, Lenkurt Demodulator, Dec. 1969.
15. *PCM Up-date Part 1*, Lenkurt Demodulator, Jan. 1975.
16. *PCM Up-date Part 2*, Lenkurt Demodulator, Feb. 1975.
17. SCHARTZ, M., *Information Transmission, Modulation and Noise*, Chapt. 4, McGraw-Hill, New York, 1959.

Chapter 5

ERRORS AND UNCERTAINTIES IN STRAIN MEASUREMENT

Vickers Shipbuilding and Engineering Ltd,
Barrow-in-Furness, UK

NOTATION

G	modulus of rigidity (shear modulus)
E	modulus of elasticity (Young's modulus)
v	Poisson's ratio
K	gauge factor
$\% \, Kt$	gauge transverse sensitivity
csa	cross-sectional area.

INTRODUCTION

The previous chapters have dealt with the factors associated with the mechanics of bonding electrical resistance strain (ERS) gauges in a variety of situations, and the instrumentation techniques that are employed to obtain strain measurements. However, in common with all types of measurement, one vital question must be asked, 'How credible is the obtained data?' Or, put another way, 'Do we satisfy Professor Peter Stein's philosophy and make valid measurements on purpose?'.

Before considering some of the principal sources of error in strain measurement, it is necessary to define the various terms that will be used. Unfortunately there is considerable confusion on this elementary aspect. For example, 'error' is often used as though it were synonymous with 'uncertainty' and 'repeatability' is often confused with 'accuracy'. Some of

209

the confusion arises because different countries often have different words for the same concept, which causes translation difficulties.

In the UK, the relevant standard is BS 5233—*Glossary of Terms used in Metrology*.[3] However, this standard is by no means totally accepted, as is evidenced by the profusion of words written on the subject.[1-6] Therefore, in this chapter, the author, using BS 5233 as a basis, understands and applies the following meanings to the general metrology terms used in connection with strain measurement.

Error (of measurement):	any departure from an accepted standard due to incomplete knowledge, or imperfection in any part of the measurement system.
Uncertainty (of measurement):	that part of the quantitative expression of the results of a measurement that states (perhaps with a specified probability) the range of values within which the true value is estimated to lie.
Repeatability:	the 'closeness' of agreement between a series of measurements made at the same time, i.e. all the readings are the *same* value.
Precision:	the repeatability of a number of individual readings about their mean value.
Accuracy:	the 'closeness' of a series of precise measurements to an accepted standard, i.e. all readings the *same* and *right*.
Reproducibility:	the 'closeness' of agreement between two or more measurements of the same quantity taken at different times.
Scatter:	the deviation from a mean value of precise readings.
Reliability:	confidence in a measurement which may be affected by uncontrollable random factors.
Resolution:	the ability of a measurement system (which includes the observer) to discriminate between two nearly equal values.

Two important aspects emerge: (1) Is the data obtained repeatable? (2) If so, is it accurate? From the above it follows that for good accuracy there *must* be good repeatability *but* good repeatability does not necessarily ensure good accuracy.

In terms of metrology, the verification or otherwise of the accuracy of a measurement system is achieved by calibration. A fundamental aspect of calibration is the traceability of an unbroken chain of valid calibration steps from the measuring device back to the relevant National Standards.

When the concept of traceability is applied to strain measurements made using bonded resistance strain gauges, certain problems are posed. The main criterion is the calibration method employed. It has been shown[7] that when a strain measurement system used to determine stress or, for example, force, is calibrated by any method other than by the repeated application of a known, traceable force, which results in a measurable reaction (which is repeatable), certain breaks in the traceability chain are inevitable. The implications and effects of this are discussed more fully later.

Consideration will now be given to the errors and uncertainties associated with a strain measurement system. These will be discussed under the following headings:

(1) Error sources.
(2) Error categories.
(3) Error avoidance and compensation.
(4) Calibration methods.
(5) Error effects, magnitudes and corrections.
(6) Errors in instrumentation.

ERROR SOURCES

To conveniently identify most of the principal sources of error, the strain measuring system will be divided into basic sections, commencing with the gauge itself and proceeding to data processing; this progression also indicates the growth of uncertainty in the measurement system as the effects of potential errors accumulate.

(a)

Gauge system and its environment	Gauge
	Adhesive
	Protection
	Environment

(b)

Signal acquisition	Joints
	Connections
	Leads

(c)   (d)

System readout	Bridge completion		Data processing	Gauge position
	Bridge supply			Calibration
	Signal conditioning			Material constants
	Electrical noise			Calculation

As a further aid, the error sources associated with individual items are tabulated below (Tables 1, 2, 3 and 4).

TABLE 1

ERROR SOURCES IN GAUGE SYSTEM AND ITS ENVIRONMENT

Gauge	(a) Change in quoted gauge factor due to large strains.
	(b) Linearity.
	(c) Tolerance on quoted gauge factor.
	(d) Exceeding quoted fatigue life.
	(e) Transverse sensitivity.
	(f) Mismatch of gauge STC and structure.
	(g) Gauge length/strain gradient ratio.
	(h) Self-heating effect.
	(i) Ratio distances from structure neutral axis to structure surface and gauge axis.
	(j) Disturbance of strain field by gauge.
	(k) Magnetoresistance effect.
	(l) Resistance change repeatability (e.g. with cyclic strain).
Adhesive	(a) Exceeding elongation capabilities.
	(b) Creep.
	(c) Hysteresis.
	(d) Insufficient cure; incorrect glue-line, inclusions and voids affecting strain transmission characteristics.
Protection	(a) Coating attacking gauge grid and soldered joints.
	(b) Breakdown permitting grid etching and causing resistive and capacitive shunting paths.
	(c) Stiffening.
Environment	(a) Effects of water, acid, corrosive substances, etc. on coating.
	(b) Effects of pressure, solar and nuclear radiation.
	(c) Effects of electromagnetic and electrostatic fields.
	(d) Effects of thermal shock.
	(e) Temperature-induced output (apparent strain).
	(f) Gauge factor variation with temperature.

TABLE 2
ERROR SOURCES IN SIGNAL ACQUISITION

Joints	(a) Attack of soldered joints by active fluxes.
	(b) Ageing of solder.
	(c) Unsatisfactory joint, e.g. 'dry joint'.
	(d) 'Tin disease'.
	(e) Generation of thermal EMFs.
Connections	(a) Change in contact resistance causing change in gauge factor desensitisation.
	(b) Effect of contact resistance change within bridge circuit.
	(c) Generation of thermal EMFs.
Leads	(a) Lead resistance causing gauge factor desensitisation.
	(b) Change in lead resistance with temperature.
	(c) Generation of thermal EMFs.
	(d) 'Noise' pick-up.

TABLE 3
ERROR SOURCES IN SYSTEM READOUT

Bridge completion	(a) Change in resistance values with temperature.
	(b) Effects of switching within the bridge circuit.
	(c) Linearity of bridge configuration.
Bridge supply	(a) Stability in out-of-balance mode of operation.
	(b) Resolution of adjustment for gauge factor.
	(c) Regulation in constant voltage or current modes.
Signal conditioning	(a) Shunting effect of amplifier input and readout device.
	(b) Resolution and linearity of readout.
	(c) Drift and linearity of amplifier.
Electrical noise	(a) Switching or transmission noise.
	(b) Series and common mode voltages and rejection ratio.
	(c) Pick-up in electromagnetic and electrostatic fields.

ERROR CATEGORIES

Errors can be categorised as follows:

(1) identifiable and avoidable;
(2) identifiable and either compensated for or corrected for;
(3) identifiable, uncorrected, but accounted for in the quoted uncertainty value;
(4) unidentifiable, unknown but, hopefully, insignificant.

TABLE 4
ERROR SOURCES IN DATA PROCESSING

Gauge position	(a) Reduction or increase of recorded strain because of gauge misalignment.
	(b) Incorrect data because of gauge mislocation.
	(c) Non-cancellation of unwanted signals (e.g. bending effects).
	(d) Grid relationship for two and three gauge rosettes.
	(e) Strain gradients parallel to the plane of the gauge.
	(f) Strain gradients normal to the test surface.
Calibration	(a) Effect of using incorrect gauge factor to compute calibration resistor.
	(b) Effect on calibration resistor of assuming nominal gauge resistance.
	(c) Effects of lead length and single arm shunting for full bridge calibration.
Material constants	(a) Difference between test material and structure constants affecting gauge factor.
	(b) Effect of material adiabatic temperature changes.
	(c) Effects of residual strain and work-hardening history.
Calculation	(a) Use of single nominal gauge factor for two and three grid rosettes.
	(b) Assumption of homogeneous material for values of E, v and ∞.
	(c) Assumption of measurement within elastic limit.

Referring to category (1), a number of potential error sources, both significant and insignificant, can be, and invariably are, introduced into the measurement system by the user with the materials and techniques that are employed. Good examples of this are (a) the use of a protective coating which liberates acetic acid, causing etching of the gauge foil, and (b) the non-cancellation of bending strains by the use of an incorrect Wheatstone bridge configuration.

When an error source produces unavoidable but known repeatable errors, i.e. category (2), it is often possible to reduce the effects, either prior to the measurements, e.g. by adjusting the calibration value to accommodate the desensitisation of the gauge factor by the lead resistance, or, alternatively, by post measurement corrections, e.g. correction of data for the different gauge factors of a three-grid rosette gauge when a common, fixed, bridge excitation has been employed. In certain circumstances, e.g. transducer manufacture, it is possible to employ certain specialised compensation techniques. This aspect of error avoidance and compensation is discussed more fully later.

Categories (3) and (4), however, present greater problems as they are less easy to define and are, therefore, the major contributors to the system uncertainty value. The situation is further complicated by the fact that there are basically two types of error, namely (a) absolute—affecting the overall accuracy, and (b) random—affecting repeatability (and therefore accuracy). Generally, random errors are easier to detect as they manifest themselves in the inability of the measurement system to give repeatable data for identical conditions, which is evident from scatter in the readings. Although random errors are readily detectable, they are difficult to predict and are peculiar to each individual measurement system. Every theoretical error analysis of a measurement system, used primarily to determine the absolute errors, should be accompanied by a series of readings obtained from the measurement system, to establish the presence, or otherwise, of any random errors in that system.

ERROR AVOIDANCE AND COMPENSATION

There are a number of error sources inherent in the technique of strain measurement using ERS gauges: detailed consideration has already been given to the importance of using recognised strain gauge materials (Chapter 1), but it cannot be over emphasised that the use of non-strain gauge materials, e.g. certain adhesives and solders, can result in large errors from, or even complete failure of, the strain measurement system. Perhaps a more important aspect is the way in which strain gauge materials are used. It is true to say that the correct use of inferior strain gauge materials can produce more credible data that the incorrect use of high quality materials. Therefore, the greatest potential source for introducing errors is the user.

Owing to the nature of strain measurement techniques, the individual skills of the operator and adherence to recognised and proven practices, within the environmental restraints, are fundamental to the credibility of the data obtained. Unless a measurement system is installed with a high standard of technique, skill, patience and self-discipline, large user-induced errors will result. Good gauging practice, using high quality materials, and an awareness on the part of the user of the consequences of the lack of attention to detail and the effects of short cuts, are the keys to the avoidance of many potential error sources.

Examples of error sources in a strain measurement system, which are often the result of inferior materials and/or workmanship are:

(a) incorrect selection of gauge, adhesive, solder, lead wire, cable and protection material;

(b) inadequate surface preparation for, and unsatisfactory application of, adhesive and protective coating—affecting bond characteristics;

(c) unsuitable soldering equipment and techniques;

(d) adhesive—incomplete cure, incorrect glue-line, inclusions and voids which affect strain transmission properties;

(e) unsuitable protection materials and techniques;

(f) moisture ingress causing (i) possible adhesive swelling, (ii) reduction of insulation 'resistance-to-ground' value, (iii) gauge grid and lead wire resistance changes;

(g) change in solder joint resistance due to (i) attack by active fluxes, (ii) ageing of solder, (iii) 'tin disease', (iv) unsatisfactory joint, e.g. 'dry joint';

(h) incorrect positioning and orientating of strain gauges;

(i) non-symmetrical cancellation of effects of transverse sensitivity, bending, temperature and other environmental conditions;

(j) changes in resistance, after initial balance, other than strain-induced gauge resistance changes.

Error sources associated with bridge excitation and balancing, calibration and signal conditioning and readout can include:

(a) incorrect selection of bridge configuration, completion components, and calibration and balancing methods;

(b) calculation errors due to changes in assumed constants, E, v, K, $\% Kt$ and elastic element dimensions;

(c) bridge excitation—type, value, self-heating and stability;

(d) thermal EMFs with d.c. excitation;

(e) induced electrical noise (i) triboelectric, (ii) magnetoresistive, (iii) electromagnetic, (iv) electrostatic;

(f) non-reproducibility of balancing point, system sensitivity and electrical calibration when conditioning and/or readout is not permanently installed;

(g) amplifier—zero and system gain stability, inferior common and series mode rejection;

(h) effects of temperature and time on system sensitivity;

(i) creep of the elastic element, gauge backing, gauge foil or gauge adhesive.

As previously stated, when an error source produces an unavoidable but known repeatable error (category (2)), compensation techniques may be employed to reduce the effect to a lower more acceptable level. This aspect is covered in depth in reference 8, and the following is a summary. It must be emphasised again that there are no satisfactory compensation techniques for errors resulting from the incorrect selection and application of strain measurement materials.

Error compensation can be divided broadly into two groups:

(1) Non-adjustable compensation—associated with the way in which the gauges etc. are used: any remaining error cannot afterwards be cancelled and therefore contributes to the total system uncertainty;

(2) Adjustable compensation—this may be employed to reduce the effects of temperature on gauge and material constants, or to standardise the individual characteristics of a number of similar measurement systems, e.g. transducers.

Non-adjustable Compensation Techniques

The most important consideration in non-adjustable error compensation is undoubtedly associated with the choice of Wheatstone bridge configuration. The use of an incorrect configuration for a particular application can result in the introduction of unnecessary error sources, the errors from which can be large and of either sign. Figure 1 illustrates the seven bridge configurations in common use. The main considerations in the selection of the most suitable configuration for a particular application can be grouped under the four main requirements:

(1) bridge output level;
(2) bridge output linearity;
(3) temperature compensation;
(4) separation of unwanted signals.

Bridge Output Levels

There are a number of techniques available for obtaining the maximum representative output from optimally positioned gauges. Probably the most common method is to increase the number of gauges used and arrange them in a Wheatstone bridge circuit. Figure 2 shows the electrical arrangement.

When this circuit configuration is energised with a constant voltage (CV) supply and used in the 'out-of-balance' mode then the output voltage is

Bridge/strain arrangement[a]	Description	Output equation $-\dfrac{V_{\text{in}}}{V_{\text{out}}}$ in mV/V	Actual strain ε Indicated strain $\hat{\varepsilon}$	Comments
	Single active gauge in uniaxial tension or compression	$\dfrac{V_{\text{out}}}{V_{\text{in}}} = \dfrac{K\varepsilon \times 10^{-3}}{4 + 2K\varepsilon \times 10^{-6}}$	$\dfrac{\varepsilon}{\hat{\varepsilon}} = 1 + \dfrac{K\hat{\varepsilon} \times 10^{-6}}{2 - K\hat{\varepsilon} \times 10^{-6}}$	Non-linear
	Two active gauges in uniaxial stress field one aligned with maximum principal strain, one 'Poisson' gauge.	$\dfrac{V_{\text{out}}}{V_{\text{in}}} = \dfrac{K\varepsilon(1 + v) \times 10^{-3}}{4 + 2K\varepsilon(1 - v) \times 10^{-6}}$	$\dfrac{\varepsilon}{\hat{\varepsilon}} = 1 + \dfrac{K\hat{\varepsilon}(1 - v) \times 10^{-6}}{2 - K\hat{\varepsilon}(1 - v) \times 10^{-6}}$	Non-linear
	Two active gauges with equal and opposite strains—typical of bending-beam arrangement	$\dfrac{V_{\text{out}}}{V_{\text{in}}} = \dfrac{K\varepsilon}{2} \times 10^{-3}$	$\dfrac{\varepsilon}{\hat{\varepsilon}} = 1$	Linear
	Two active gauges with equal strains of same sign—used on opposite sides of column with low temperature gradient (bending cancellation, for instance).	$\dfrac{V_{\text{out}}}{V_{\text{in}}} = \dfrac{K\varepsilon \times 10^{-3}}{2 + K\varepsilon \times 10^{-6}}$	$\dfrac{\varepsilon}{\hat{\varepsilon}} = 1 + \dfrac{K\hat{\varepsilon} \times 10^{-6}}{2 - K\hat{\varepsilon} \times 10^{-6}}$	Non-linear

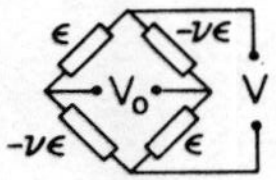	Four active gauges in uniaxial stress field two aligned with maximum principal strain, two 'Poisson' gauges (column).	$\dfrac{V_{\text{out}}}{V_{\text{in}}} = \dfrac{K\varepsilon(1+v) \times 10^{-3}}{2 + K\varepsilon(1-v) \times 10^{-6}}$ $\dfrac{\varepsilon}{\hat{\varepsilon}} = 1 + \dfrac{K\hat{\varepsilon}(1-v) \times 10^{-6}}{2 - K\hat{\varepsilon}(1-v) \times 10^{-6}}$	Non-linear
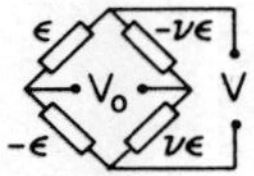	Four active gauges in uniaxial stress field—two aligned with maximum principal strain, two 'Poisson' gauges (beam).	$\dfrac{V_{\text{out}}}{V_{\text{in}}} = \dfrac{K\varepsilon(1+v) \times 10^{-3}}{2}$ $\dfrac{\varepsilon}{\hat{\varepsilon}} = 1$	Linear
	Four active gauges with pairs subjected to equal and opposite strains (beam in bending or shaft in torsion).	$\dfrac{V_{\text{out}}}{V_{\text{in}}} = K\varepsilon \times 10^{-3}$ $\dfrac{\varepsilon}{\hat{\varepsilon}} = 1$	Linear

[a] $(R_1/R_4)_{nom} = 1$; $(R_2/R_3)_{nom} = 1$ when two or less active arms are used.

FIG. 1. Wheatstone bridge non-linearity. Constant voltage power supply is assumed, ε and $\hat{\varepsilon}$ (strains) are expressed in microstrain units (in/in $\times 10^6$). Courtesy of Micro Measurements.

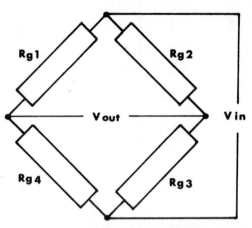

Fig. 2.   Basic Wheatstone bridge circuit.

directly related to surface strain. It can be shown that, when $R_{g1} = R_{g2} = R_{g3} = R_{g4}$, the bridge output, V_{out}, for a single active arm is

$$V_{out} = \frac{K \times \varepsilon}{4 + 2K \times \varepsilon} V_{in}$$

where $K = (\Delta R_g / R_g \varepsilon)$.

When the resistance of R_{g1} and R_{g3} increases, and that of R_{g2} and R_{g4} decreases by the same amount, then the output will be four times that obtained when R_{g1} only changes by that amount. This is termed a 'full bridge' with four 'active arms'. In practice, this would be achieved by R_{g1} and R_{g3} being subject to, say, tensile surface strain and R_{g2} and R_{g4} to compressive strain.

The signal levels obtained from a Wheatstone bridge are normally low, but there are a number of ways of increasing the effective output. One common method, which is often misused, is to increase the bridge excitation voltage. The level to which the bridge voltage can be raised is governed by the heat dissipation of the gauge, which is a function of grid area, resistance and the 'heat sink' characteristics of the material to which the gauge is bonded. Recommended levels of bridge excitation are clearly laid down by the gauge manufacturer, with the corresponding stability figures.[9] A dissipation level of $1 \cdot 5 \, mW/mm^2$ of grid area is a typical value for high stability and $7 \cdot 5 \, mW/mm^2$ will produce reasonably good static stability. Higher resistance gauges, or several gauges in each arm of the bridge, enable higher levels of excitation to be used, thereby increasing the relative output without degrading stability.

Another method of increasing the effective output is to employ an asymmetrical bridge configuration. It can be shown that when a symmetrical bridge (i.e. $R_{g1} = R_{g2} = R_{g3} = R_{g4}$) is employed, only 50% of the possible maximum output is obtained.[10] By employing an asymmetrical bridge technique (i.e. $R_{g1} = R_{g2} \neq R_{g3} = R_{g4}$), increases of up to 80% of the maximum are practicable, with the same dissipation in the gauge. However, this technique is limited to $\frac{1}{4}$ and $\frac{1}{2}$ bridge configurations.

When a Wheatstone bridge is used in the other possible configuration i.e. 'null balance', the excitation voltage does not affect the output, it only affects the sensitivity of the null indicator. This makes null balance systems very suitable for battery operated instruments.

One further practicable method of increasing the bridge output, especially for transient measurements, is to pulse the bridge excitation at levels up to 80 times higher than the normal d.c. level.[11]

Bridge Output Linearity

It will be noted from Fig. 1 that only certain bridge configurations produce a linear output for a linear surface strain effect. Indeed, the only configurations that do produce a completely linear output are those employing bridges with equal and opposite arms, i.e. 2 fully active arms and 4 fully active arms. However, the error due to non-linearity, when present, is normally small, and can usually be ignored when measuring elastic strains in metals. The percentage non-linearity error increases with the magnitude of the strain being measured, and can become quite significant at large strains (for example, the error is about 0·1% at 1000 $\mu\varepsilon$, 1% at 10 000 $\mu\varepsilon$, and 10% at 100 000 $\mu\varepsilon$; or, as a convenient rule of thumb, the error, in percent, is approximately equal to the strain, in percent).[12]

The character of the non-linearity associated with the $\frac{1}{4}$ bridge configuration can be illustrated by expressing the bridge output equation in the following form:

$$\frac{V_{out}}{V_{in}} = \frac{K\varepsilon \times 10^{-3}}{4} \left(\frac{2}{2 + K\varepsilon \times 10^{-6}} \right)$$

where V_{out}/V_{in} = dimensionless bridge output (mV/V); V_{out} = output voltage (mV); V_{in} = bridge supply voltage (V); K = gauge factor of strain gauge; and ε = strain ($\mu\varepsilon$, microstrain).

In the above equation, the term in parentheses represents the non-linearity. It is evident from the form of the non-linearity term that its magnitude will be less than unity for tensile strains and greater than unity for compressive strains, and the errors in strain indication due to the non-linearity will be of the same nature. In other words, indicated tensile strains will be too small and indicated compressive strains too large.

Temperature Compensation

The introduction of the self-temperature-compensated (STC) gauge has resulted in a marked improvement in compensation techniques for that undesirable characteristic of the metal foil gauge, its resistance change with temperature. The STC gauge was primarily developed to eliminate the need for dummy gauges, simplifying the use of the single active gauge and enabling more practical measurements with rosette type gauges. But care must be exercised in the use of STC gauges, in particular the choice of bridge configuration, $\frac{1}{4}$ bridge being the most critical.

In a large number of applications, when a single active arm is employed, the bridge completion is remote from the gauge (see Fig. 3). This arrangement has two undesirable effects. The first is a fixed reduction in system sensitivity (because of added series resistance) known as 'de-sensitisation', and the second is the change in lead and plug and socket contact resistance (ΔR_L and ΔR_C) because of changes in temperature of the leads and connectors.

Desensitisation can be allowed for by modifying the gauge factor, i.e.

$$K' = \frac{R_g}{R_g + R_{L1} + R_{L2} + R_{C1} + R_{C2}} K$$

where K is the quoted gauge factor and K' the modified value.

However, the effects of ΔR_L and ΔR_C can still produce considerable errors, but these effects can be considerably reduced by employing a 'three-wire system' (Fig. 4). This arrangement transfers the bridge apex point to

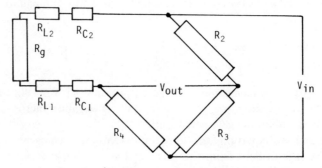

FIG. 3. $\frac{1}{4}$ bridge two-wire configuration.

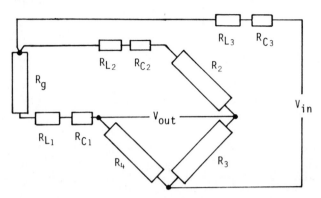

FIG. 4. $\frac{1}{4}$ bridge three-wire configuration.

the gauge, removing the elements R_{L2} and R_{C2} from being in series with R_g and putting them in series with R_2. For this method to be fully effective the resistance of R_g must be equal to that of R_2. The balance equation has to be modified to

$$(R_g + R_{L1} + \Delta R_{L1} + R_{C1} + \Delta R_{C1})R_3$$
$$= (R_2 + R_{L2} + \Delta R_{L2} + R_{C2} + \Delta R_{C2})\acute{R}_4$$

It will be noted that the effect of R_L, R_C and the temperature and vibration changes induced in them is now the *difference* between R_{L1} and R_{L2} etc., rather than the *sum* as in the two-wire case. Also, the system desensitisation is now reduced by half, i.e.

$$K' = \frac{R_g}{R_g + R_{L1} + R_{C1}} K$$

The effects of R_L, R_C, ΔR_L and ΔR_C can be further reduced by using the highest value of R_g possible (Table 5).

The three-wire system *must* be used when using STC gauges for static strain measurements under variable ambient temperatures, and in situations where the components of the bridge are at different temperatures. For the three-wire system to be effective, the following conditions for the two leads etc., within the bridge (i.e. R_{L1}/R_{C1} and R_{L2}/R_{C2}) must be fulfilled: (1) they must have the same resistance and temperature coefficient of resistance, be of the same thermal mass and experience the same thermal conditions, and (2) they must be in series with the same resistance value.

Although the three-wire system permits a considerable reduction in the effects of leads and connectors, a small secondary effect is introduced in the

TABLE 5
LEAD EFFECTS IN TWO- AND THREE-WIRE CONNECTIONS

	Two-wire both lead lengths	Three-wire difference in length	Two-wire (both lead values)		Three-wire (lead difference values)	
	$(R_{L1} - (R_{L2})$	$(R_{L1} + (R_{L2})$	120	350	120	350
Temperature	Jumper leads	1/020	$17\ \mu\varepsilon/°C/m$	$5\cdot8\ \mu\varepsilon/°C/m$	$8\cdot5\ \mu\varepsilon/°C/m$	$2\cdot9\ \mu\varepsilon/°C/m$
	Signal cable	16/020	$1\cdot3\ \mu\varepsilon/°C/m$	$0\cdot46\ \mu\varepsilon/°C/m$	$0\cdot65\ \mu\varepsilon/°C/m$	$0\cdot23\ \mu\varepsilon/°C/m$
			Two lead values		*One lead values*	
Gauge factor desensitisation (at constant temperature)	Signal cable	16/020	$0\cdot05\ \%/m$	$0\cdot025\ \%/m$	$0\cdot02\ \%/m$	$0\cdot01\ \%/m$

form of the third lead, i.e. R_{L3}, R_{C3} (Fig. 4). The user often has the choice as to whether this lead is placed in the bridge input (V_{in}) or the bridge output (V_{out}) when using a symmetrical bridge. The basic theoretical considerations are: the relationship, in the input case, of the lead resistance to bridge input resistance—affecting the circuit sensitivity; and, in the output case, the relationship of the lead resistance to detector sensitivity— affecting the load resistance as seen by the bridge. In a number of cases where the third wire passes through high magnetic fields, the induced series mode signal is often large compared with the signal being detected, causing large errors where the third wire is connected in the bridge output, and the detector series-mode rejection is inadequate. Consequently, the third wire in Fig. 4 is shown in the V_{in} circuit form. The third wire need not fully conform to condition (1) above, but it is preferable that it experiences the same thermal conditions as the two signal leads.

If an asymmetrical bridge is employed (e.g. $R_{g1} = R_{g2} \neq R_{g3} = R_{g4}$), with the asymmetry across the supply, the user has no choice regarding the position of the third wire as it *must* be placed in the supply lead for the circuit to conform to condition (2).[10] A choice of lead position is also available for a $\frac{1}{2}$ bridge configuration and the same considerations apply, the lead in the supply line being preferable, to reduce the possibility of any series-mode pick-up.

The STC gauge does not completely eliminate the temperature-dependent resistance changes of the gauge foil, but does provide a degree of compensation for it, related to a particular structure material. This is achieved by the manufacturer modifying the unbonded foil resistance change characteristic due to temperature, to the inverse of the difference between the expansion coefficient of the material for which the gauge is intended and the expansion coefficient of the gauge. An example for steel is given in Fig. 5, which shows the net effect of the two characteristics when the gauge is bonded to the material. There are inevitable differences in the STC characteristic between individual gauges of the same batch and between different batches. Typical spreads from 10 gauges from the same batch are illustrated in Fig. 6. Therefore, when $\frac{1}{2}$ bridge and full bridge configurations are employed, gauges of the same batch should be used to reduce the STC gauge temperature mismatch.

Separation of Unwanted Signals

In a number of applications, it is often necessary to separate the required strain signal from other unwanted signals which may also be present. Probably the most common case is the separation of, and therefore

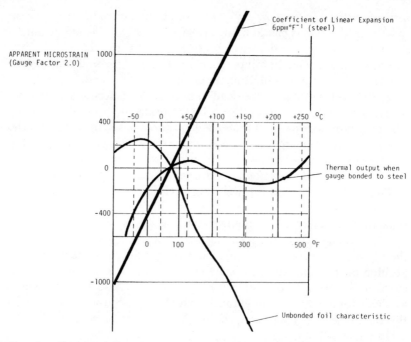

FIG. 5. Characteristics of gauge with self-temperature compensation for steel.

compensation for, the effects of bending strains. This is usually achieved by arranging the active gauges within the bridge such that, for example, gauges R_{g1} and R_{g3} (Fig. 2) will experience the same sign and magnitude 'required signal' strains but the opposite sign and magnitude bending strains, thus effectively cancelling out the bending strain component. A good example of this is the commonly used 'tension link' transducer illustrated in Fig. 7. Nevertheless this is not as simple as it may first appear, particularly if the tension link experiences any rotation.

Consider a tension link and the use of two twin-grid gauge patterns (of the same batch and with each grid type having identical characteristics) placed accurately circumferentially opposite, then, to produce a neat installation and facilitate equal lengths within the bridge circuit, the natural tendency is to arrange the gauges according to Fig. 8(a). However, if the gauge backing centre-line is not exactly on a bending strain node then the axial grids in the two opposite bridge arms will be experiencing opposite but not equal magnitudes, because of transverse misalignment. The size of the error produced will depend upon factors such as bending strain gradient,

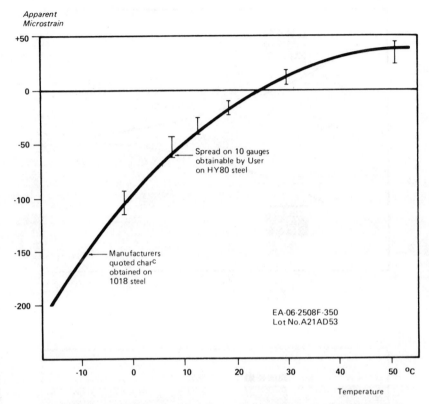

FIG. 6.   Reproducibility of quoted thermal output characteristics ($K = 2 \cdot 0$).

change in bending strains after bridge balancing, and the nature of bending strains in the elastic member. Better compensation is obtained when gauges are positioned as shown in Fig. 8(b) (although this may mean a less neat installation).

Another example of bending compensation using the bridge configuration is the determination of stresses in large plates that are also subject to local bending. In this application the position of the gauges in the

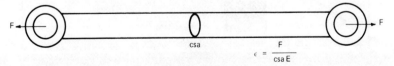

FIG. 7.   Tension link.

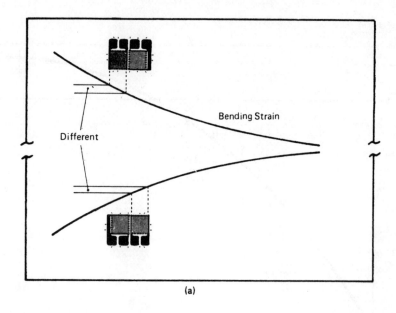

(a)

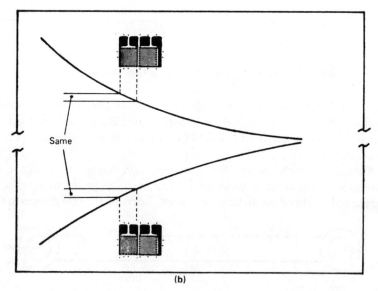

(b)

FIG. 8.   Tension link pure bending compensation.

stress field and the arrangement of the number of active gauges and their relative positions in the bridge, perform a mathematical function. When only one side of the plate is accessible, it is possible to measure strains with gauges raised above the plate at distances of $\frac{1}{2}$ plate thickness (ε_s) and plate thickness (ε_r) using a mechanical bridge, or a separator which responds to the bending curvature of the plate. It can be shown[13] that ε_a the axial strain using a 'double-decker' system is given by

$$\varepsilon_a = 3\varepsilon_s - 2\varepsilon_r$$

When the five gauges are arranged in a bridge as shown in Fig. 9, with three fixed resistors (R_f), or dummy gauges for temperature compensation, the bridge output will be mathematically related to ε_a. It must be remembered though, that several gauges in each arm will produce an average of the surface strain affecting the gauges, which may not be desirable, particularly in high strain gradient situations.

As indicated, where practicable and necessary for the particular strain measurement requirement, a number of error sources can be compensated for by the use of a fully active Wheatstone bridge, and the particular arrangement of active gauges within the bridge. However, a good example of an error for which there is no compensation technique is that due to induced bending strains caused by misalignment of the gauge itself. The implications of gauge misalignment are not always fully appreciated and

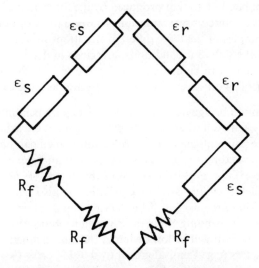

FIG. 9.   'Double-decker' bridge arrangement.

further consideration of a tension link will illustrate the problem. Circumferential and transverse misalignment has already been considered but by far the most important consideration is the complex situation which results from axial misalignment of the gauges and the effect this has on bending strain compensation. Considering two twin-grid gauges placed circumferentially opposite on a circular elastic element that is subject to pure bending, four different situations occur as a result of axial misalignment. These are:

(1)   Both pairs misaligned by the same amount with the induced bending strain either of the same sign as, or of opposite sign to, that of the axial strain.

(2)   One pair axially aligned and the other misaligned with the induced bending strain of the same sign as, or of opposite sign to, that of the axial strain.

(3)   Both pairs misaligned by different amounts with the induced bending strains in each pair either of the same sign as, or of opposite sign to, that of the axial strain, or the reverse of this.

(4)   Both pairs axially aligned.

For each of the eight different conditions outlined in (1) to (3) above, there is another major contributory factor which affects the magnitude of the error due to axial misalignment; that is the ratio of the surface strain resulting from bending to that produced by the force acting upon the link. A full analysis of this complex situation is beyond the scope of this chapter. However, examples of the three different situations in which error occurs, with some of the different conditions, are given in Table 6. A number of general statements can be made regarding the relationship between axial misalignment of two twin-grid gauges and pure bending cancellation.

(1)   Maximum bridge output with complete pure bending cancellation is only achieved when (a) both pairs of gauges are axially aligned or (b) one pair is aligned and the other misaligned pair have induced in them a bending strain of *equal* magnitude and *opposite* sign to that of the axial strain, irrespective of the degree of misalignment.

(2)   Complete bending strain cancellation is also achieved when (a) both pairs are misaligned by the same amount, irrespective of the axial strain/bending strain ratio and sign relationship, although the bridge output will be low owing to the misalignment of the gauges themselves (e.g. 1° misalignment—0·061% error (see Table 6); (b) one pair is aligned and the other misaligned pair have induced in

TABLE 6

PERCENTAGE ERRORS PRODUCED IN 2·6 ACTIVE ARM TENSION LINK TRANSDUCER DUE TO INDUCED PURE BENDING STRAINS RESULTING FROM GAUGE MISALIGNMENT

Condition	Both pairs misaligned by same amount as indicated — Axial strain/bending strain ratio same sign and opposite sign			One pair aligned—other pair misaligned as indicated						Example: both pairs misaligned as indicated with a fixed misalignment of 5° between each pair e.g. 0·5° = 5·0° and 5·5°					
				Same sign (0°)			Opposite sign (180°)			Same sign (0°) +ve			Opposite sign (180°) −ve		
	10:1	1:1	1:10	10:1	1:1	1:10	10:1	1:1	1:10	10:1	1:1	1:10	10:1	1:1	1:10
Angle of misalignment															
0·5	−0·015	−0·015	−0·015	−0·008	−0·015	−0·084	−0·007	0	+0·069	−1·017	−1·837	−10·036	−0·835	−0·015	+8·184
1°	−0·061	−0·061	−0·061	−0·034	−0·061	−0·335	−0·027	0	+0·274	−1·229	−2·185	−12·090	−1·017	−0·061	+9·498
2°	−0·244	−0·244	−0·244	−0·134	−0·244	−1·340	−0·110	0	+1·096	−1·743	−2·970	−15·241	−1·417	−0·244	+12·027
3°	−0·548	−0·548	−0·548	−0·301	−0·548	−3·013	−0·247	0	+2·465	−2·371	−3·874	−18·841	−2·045	−0·548	+14·419
4°	−0·973	−0·973	−0·973	−0·535	−0·973	−5·352	−0·438	0	+4·379	−3·130	−4·894	−22·539	−2·738	−0·973	+16·672
5°	−1·519	−1·519	−1·519	−0·836	−1·519	−8·356	−0·684	0	+6·837	−4·001	−6·031	−26·332	−3·549	−1·519	+18·782

them a bending strain of *equal* magnitude to and the *same* sign as that of the axial strain, irrespective of the degree of misalignment—again the bridge output will be low; (c) initially, both pairs misaligned by the same fixed amount, and the induced bending strain of *equal* magnitude to and *opposite* sign from that of the axial strain in one pair which is further misaligned, irrespective of the degree of that misalignment—bridge output will again be low.

It will be noted from Table 6 that for all other conditions, each case is unique, non-linear and, in the condition where one pair is aligned and the other misaligned such that the induced bending strain is of opposite sign to, and of greater magnitude than, the axial strain, a positive error is produced (i.e. a case where gauge misalignment results in an *increase* in bridge output). This condition is illustrated in Fig. 10 for the even more complex situation where the bending strain is cyclic because of rotation of the measurement system (e.g. when measuring thrust in a propeller shaft). Figure 10 illustrates a 5° misalignment between 0° (same sign) and 180° (opposite sign) (see bottom line, middle case, Table 6) and clearly shows that an induced sinusoidal bending strain results in an output which can be anywhere in the 'envelope' (i.e. $-ve$, or $-ve$ and $+ve$) depending upon the axial/bending strain ratio. Even though it is a once-a-cycle event, it is difficult to identify and apply a correction for the error produced, especially when the bending strain/axial strain ratio changes during the cycle. As the potential error is large (i.e. $+6.8\%$ for a $\varepsilon_a/\varepsilon_b$ of $1/10$ at 5° misalignment) it could well mask a required signal, e.g. axial vibration due to the dynamic unbalance of a propeller blade. Another complication in this case is the far larger error from the torsional strain induced in the misaligned grids.

Adjustable Compensation Techniques

With regard to adjustable compensation, the main application is undoubtedly associated with strain gauge transducers.[8] However, in certain other applications some, if not all, of the techniques can produce worthwhile reductions in the effect of errors. It should be emphasised that, in the majority of cases, these techniques can only be used where the elastic element can be removed from its operating environment or a simulated environment can be produced around it to enable the adjustments to be made.

When compensating for thermal effects it is advisable to fit any protective covers or other mechanical parts, or to simulate any mass that may have an effect upon the thermal characteristics of the elastic element.

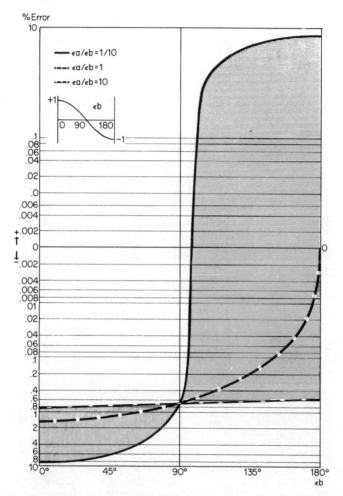

FIG. 10. Percentage errors produced in 2·6 active arm tension link transducer due to pure bending strains resulting from one pair of gauges misaligned by 5°.

The four basic parameters for which compensation can be made are listed in Table 7, with the fundamental causes of the errors and the means of reducing them by correct selection and use of materials and techniques. A more detailed consideration of the compensation techniques is given in references 14 and 15, on which the following is based; it must be emphasised, however, that most of the compensation techniques are suitable only for constant voltage bridge supplies. Also, it is essential to

TABLE 7

ERROR SOURCES AND COMPENSATION TECHNIQUES

Error	Cause	Reduction	Compensation technique
Initial bridge unbalance	Non-symmetry of adjacent arm gauges 'as-laid' resistance.	Good gauging practice.	(1) Increase gauge resistance by rubbing down grid. (2) Use bondable, adjustable constantan bridge–apex resistor.
Thermal zero-shift	Non-symmetry of temperature effects on either/both adjacent bridge arms.	Use gauges of same foil lot and production batch, and STC, on common backing. Symmetry of adjacent arm interwiring, soldered joint and thermal mass, must experience same thermal conditions.	(1) Adjust bridge interwiring. (2) Use bondable, adjustable copper bridge–apex resistor.
Thermal span-shift	Changes in temperature affecting gauge factor and modulus of elasticity of elastic element.	Use of high quality gauges and elastic element material.	(1) Use 'Karma' foil (modulus compensated) gauges. (2) Use bondable, adjustable 'Balco' resistor in series with constant voltage supply.
Individual device span characteristic adjustment	Spreads in machining tolerances of elastic element. Gauge misalignment and incomplete strain transmission. Differential in gauge characteristics.	Tight tolerances on dimensions affecting cross-sectional area. Good gauging practice with high quality materials.	(1) Use bondable, adjustable constantan resistor in series with constant voltage supply.

prepare for their use either during the gauge bonding phase (in the case of bondable resistors) or during bridge wiring (in the case of adjustment of bridge interwiring).

The decision whether or not to employ the compensation techniques indicated depends upon the accuracy required from the measurement system and is, therefore, directly related to the temperature changes the gauge installation will experience. The order in which the adjustments of the various compensation devices are made is critical, as some are interdependent.

Compensation for initial bridge unbalance is probably the easiest to achieve as no temperature or load changes are necessary. However, this adjustment should not be made until thermal stability has been achieved. Where possible, the installation should, after the bridge has been wired, be 'post-cured'. A post-cure of 1 h at 50 °C above the system operating temperature is usually adequate. The need for thermal ageing is illustrated in Fig. 11, which shows the zero shift due to thermal ageing of a 'filled' room temperature cured polyester strain gauge adhesive, which, although ideal for certain adverse environment stress analysis applications, can produce the thermal creep characteristics shown. There are three basic methods of balancing the bridge circuit to give zero output for zero load (or for any other suitable load) once thermal stability has been established. The installed gauge resistance of an open-faced gauge can be adjusted by very carefully stroking the gauge, in the direction of the grid, with a pencil eraser or a cotton applicator charged with fine pumice powder. Any residue or contamination must be removed with neutraliser. This method can, however, result in a change of the STC characteristic of the gauge. In the case of encapsulated gauges, a bondable adjustable constantan resistor wired into the bridge apex gives satisfactory results. It is possible to obtain a bridge balance by permanently wiring a high value fixed resistor across one arm in order to reduce the resistance value, although this will produce an increased non-linear characteristic from the gauge. This resistor should be a stable, low temperature coefficient (i.e. < 100 ppm/°C) component, with good ageing properties, so as not to introduce instability.

Without any compensation for temperature-induced zero shift, a wide range of false outputs is possible, depending upon the type of gauge and the patterns used, the symmetry of the thermal mass and the conditions of the adjacent bridge arms. For example, a full bridge pattern gauge can have a thermal output characteristic of, typically, 0·1 microstrain/°C, and values of up to 2·5 microstrain/°C can be expected from two twin-grid gauges. Figure 12 illustrates the thermal zero shifts obtained from four tension

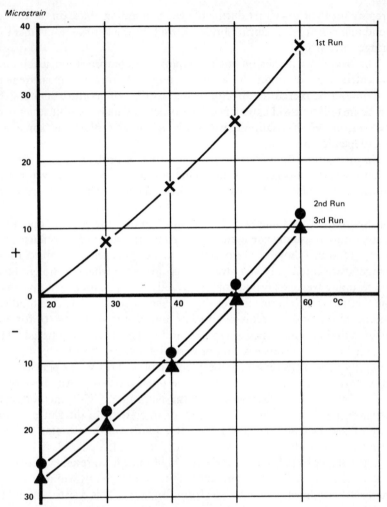

FIG. 11.   Thermal ageing characteristics.

links: note that the spreads of the four 'identical' links are considerable, even though three of the devices used gauges of the same foil lot and production batch. If, however, the techniques listed in Table 7, e.g. bondable, adjustable copper bridge–apex resistors are employed, it is not difficult to reduce the values to, typically, 0·02 microstrain/°C, provided that the elastic element material and installation are of suitable quality.

As the gauge factor and the elastic element modulus of elasticity are

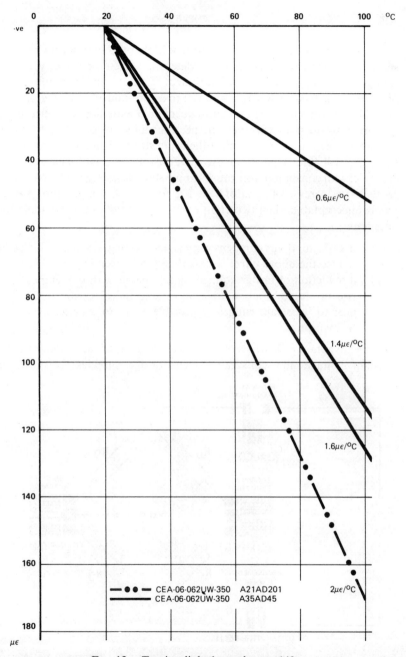

FIG. 12. Tension link thermal zero shifts.

functions of temperature, it is not surprising that the span sensitivity of a strain measurement system used to determine stress or force, will also change with temperature. If the instrument readout is in strain units, then only the gauge factor variations give rise to errors, and all other factors will normally be corrected for in the subsequent data reduction, converting the measured strain into whatever engineering units are of interest. If the instrument readout is in any units other than strain, then the magnitude of the error will also depend on the material of the elastic element. For example, with constantan foil gauges and a steel elastic element, errors can result in an increase of span of 1 to 2 % for a 55 °C temperature rise. If this is unacceptable, then two compensation techniques are available for improvement:

(1)   use of a modulus compensating gauge foil (e.g. Karma), which can reduce the above effect to less than 0·5 % over 55 °C; or

(2)   a reduction of bridge voltage, simultaneous with the temperature rise, produced by a bondable, adjustable 'Balco' resistor situated as near to the strain gauges as possible, so as to experience similar temperatures.

One big advantage of Karma foil is that compensation is achieved for both constant voltage and constant current bridge supplies. It must be

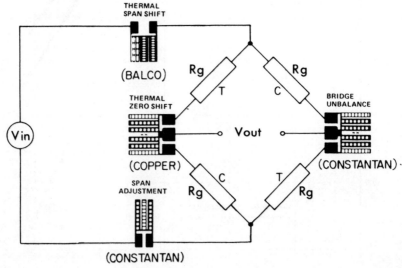

FIG. 13.   Bridge circuit with compensation resistors. Courtesy of Micro Measurements.

emphasised, however, that adjustments of span changes with varying temperatures are usually the most difficult to achieve, because the force must be applied to the elastic element at different temperatures. The practicalities of this often preclude any compensation being applied (e.g. a propeller shaft).

Compensation for span error (i.e. to produce a number of measurement systems with identical outputs) is carried out during the calibration process, after the adjustment of thermal zero and span shifts, by adjusting a temperature insensitive resistor in series with the voltage supply. However, considerable difficulties are encountered when a measurement system is calibrated by electrical simulation techniques (see Calibration Methods, later in this chapter).

Figure 13 illustrates the bondable temperature-compensated (constantan), temperature sensitive (Balco) and very low resistance (copper) adjustable resistors, for the four parameters discussed.

CALIBRATION METHODS

Implicit in the consideration of error sources and any resulting errors and uncertainties, must be the question of calibration. The distinction between repeatability and accuracy has already been drawn. Repeatability is usually achieved by taking a series of readings from the system and determining the spread of the readings, e.g. $\pm 2\%$. Accuracy, on the other hand, can only be determined by calibration against a known standard, which is traceable back to National Standards. It is, therefore, possible to have a $\pm 2\%$ repeatable system that has an accuracy of only 50%.

It is necessary to draw a distinction between the use of foil gauges as sensing elements for transducers directly calibrated against a known force and used to measure force, and their use in computing stress or force from strain measurements. In the case of the transducer, since strain is the fundamental quantity measured by strain gauges, all transducers employing them are constructed such that the physical variable (e.g. force) deforms one or more elastic members, to which gauges are bonded. In this case, the measured strain is proportional to the applied force and the gauge output can be directly related to the known force, thereby enabling a direct, traceable calibration.

In the second case, where stress or force is computed from measured strain, the application of metrology principles to the strain measurement can present major problems, depending upon the method of calibration used. Two principal calibration techniques for strain measurement are possible, (a) direct calibration with a known force (resulting in a known strain) and, usually when (a) is not possible, (b) calibration by electrical simulation.

Calibration Against a Known Force

From many aspects, direct calibration against a known applied base or derived unit for strain measurement as well as for transducers is preferable, primarily, of course, because it can be traceable to National Standards if all the relevant items of the readout system are themselves traceable. This also verifies that the effects of a number of absolute error sources have been eliminated, as their effects are reflected in the output obtained from the measurement systems in response to the applied force. These major error sources are: absolute value of gauge factor; repeatable strain transmission characteristic; non-uniformity of elastic element dimensions; local stress concentrations; uncertainty in the material constants; gauge misalignment etc.

Ideally, calibration with a known force should be carried out in a similar environment to that in which the device will be used. In the case of transducers, during calibration adjustment to the compensation technique for any non-linearity in the output can be made, together with span adjustments to provide the readout in engineering units. It must be emphasised that the effect of using a measuring system in a different environment, or in a different mode to that of the calibration, may invalidate the traceability.[16] During calibration with an applied force, the measuring system can also be checked for creep under load, having first established that the system is drift-free under no load. Errors which affect repeatability, e.g. temperature, cannot easily be checked by this calibration method.

For most calibrations against a known force, static loading is better than dynamic loading, and a digital display should be used in preference to an analogue readout as it provides higher resolution, without the potential for misinterpretation by operator subdivision of the minor scale divisions. However, where dynamic load calibration is necessary, an analogue recording (with a time scale) is essential. After the force/strain relationship has been established with the known force, it is advisable to obtain a 'check calibration' value from an electrically simulated calibration and to use this on-site to verify that the system sensitivity remains constant from the time of the direct calibration to the end of the measurement phase.

Calibration by Electrical Simulation

In marked contrast to calibration against a known traceable force, which produces a readout directly related to the cause, calibration by electrical simulation, derived from calculation, has many intermediate stages, and hence many potential error sources. These include gauge and material 'constants'; total surface strain transmission and gauge system response to it; and the integrity of the instrumentation system.

There are two basic methods of electrical simulation:

(1) injection of a voltage or current signal equivalent to the output produced by the force;

(2) simulation of a resistance change equal to the net effect of those produced in the Wheatstone bridge by the surface strain.

At first sight, injecting a voltage or current signal may seem to be a completely satisfactory method of calibration. However, it has two fundamental drawbacks; firstly, it invariably necessitates disconnection of part of the bridge circuit, and secondly, it does not check in any way that the gauge is correctly wired into the bridge or that the correct signal polarity is obtained at the output. It is therefore, not recommended.

Two 'change–resistance' methods are possible:

(1) the introduction of a low value resistance—in series with a tensile-sensing bridge arm, for a tensile strain simulation, and in series with a bridge arm adjacent to a compressive-sensing arm, for a compressive strain simulation;

(2) the introduction of a high value parallel resistance—across a compressive-sensing arm for a tensile strain simulation, and across a tensile-sensing arm for a compressive strain simulation.

The series method tends to be less accurate, because of the relatively large effects of lead and contact resistance in relation to the low value of the calibration resistance itself. The parallel method is therefore the more usual and is termed 'shunt calibration'.

When a strain measurement system is calibrated by electrical simulation, the basic assumption that the surface strain of the loaded structure is transmitted through the adhesive layer, via the gauge backing, to the gauge foil to provide a related resistance change is unproved. Indeed, any simulated method relies upon the assumption that complete strain transmission is achieved and remains linear over the complete range and that the simulated strain, based on the theoretical calculated strain, is the same as the measured strain. The validity of the data produced by a device which is calibrated by simulation methods is, at best, only as good as the assumptions made about uncertainties in the calibration.

There are a number of factors involved in theoretical calibration. Some produce relatively large errors (up to 5 %), while others, although causing relatively small individual errors (0·1 to 0·5 %), can be cumulative and therefore significant in the data obtained. All factors must be considered in order to verify the magnitudes and signs of the errors produced. It will be necessary in almost every case to calculate an individual value of shunt resistance, rather than employ special techniques[17] where adjustment of bridge excitation (effectively gauge factor) and preselected resistance values are employed. It is also advisable to perform a theoretical calibration for a number of points within the full range output to establish the linearity of the complete instrumentation system, including the readout device. To electrically simulate calibration of a strain measurement system from calculated values of resistance it is essential to know:

For strain readout:
    (i)   the gauge factor;
    (ii)   the gauge transverse sensitivity.

For stress determination:
    (i)   both the above;
    (ii)   the directions of principal strains;
    (iii) the elastic element constants e.g. E and v or G;
    (iv) the amount of gauge(s) misalignment;
    (v)   the type of stress field.

For force determination:
    (i)   all the above;
    (ii)   the csa of the structure at the gauge(s) position(s).

Using these parameters, a series of calculations, shown in Table 8, are made, which result in a value for shunt resistance, R_{sh}. Table 9 gives values of R_{sh}, with the effective change in gauge resistance for different nominal gauge resistances and active arm (AA) combinations, with the corresponding simulated microstrain for a gauge factor of 2·1. The values of R_{sh} in the table are calculated using

$$R_{sh} = \frac{R_g \times 10^6}{K \times \varepsilon_s \times AA} - R_g$$

To obtain values of R_{sh} for simulated strains up to 1500 microstrain, to within $-0·3 \%$, the equation becomes

$$R_{sh} = \frac{R_g \times 10^6}{K \times \varepsilon_s \times AA}$$

TABLE 8

SHUNT RESISTANCE CALIBRATION PATH

$$R_{sh} \to \varepsilon_s \to \varepsilon_m \to \sigma \to E \to F$$

Expression		Uncertainties— It is assumed that:
$\varepsilon_s = \dfrac{R_g \times 10^6}{(R_{sh} + R_g) \times K \times AA}$	K	Value quoted by manufacturer is valid and does not vary with time or temperature
	AA	All active arms have the expected output relationship (e.g. $1\cdot0 + 1\cdot0 + 0\cdot285 + 0\cdot285$)
	R_{sh}	Effects of lead length and switch connection resistance are negligible
$V_{out} = \dfrac{\Delta R_g}{4R_g} V_{in} \times AA$	$\dfrac{\Delta R_g}{R_g}$	Surface strain is the only cause of this ratio
	V_{in}	The supply is stable with time and temperature
	V_{out}	The output is not altered in any way by the instrumentation and readout device
$\varepsilon_m = \dfrac{4V_{out}}{V_{in} \times K \times AA}$	ε_m	Complete strain transmission is achieved and that this value is equal to ε_s
	V_{out} V_{in} K AA	(As above)
$\sigma = E(\varepsilon_m)$	E	The value is known and does not vary with time and temperature
	σ	Only axial stress is present and is within the elastic limit
$F = \sigma \,(csa)$	csa	Elastic element form known, i.e. truly circular when πR^2 is used or oval when $\max D \times \min D \times \pi/4$ is used
	σ	(As above)
Compound expression		

$$F_s = \frac{R_g}{R_g + R_{sh}} \times \frac{E \times csa \times 10^6}{K \times AA}$$

TABLE 9
SHUNT RESISTANCE CALIBRATION VALUES

%	μ Strain active arms			120Ω		350Ω		1000Ω	
	1	2	4	ΔR_g	R_{sh}	ΔR_g	R_{sh}	ΔR_g	R_{sh}
0·0001	1	$\frac{1}{2}$	$\frac{1}{4}$	0·0002	57M	0·00073	167M	0·0021	476M
0·001	10	5	$2\frac{1}{2}$	0·00252	5·7M	0·00735	16·7M	0·0210	47·6M
0·010	100	50	25	0·0252	570K	0·0735	1·67M	0·210	4·74M
0·015	150	75	$37\frac{1}{2}$	0·0378	381K	0·11025	1·11M	0·315	3·17M
0·025	250	125	$62\frac{1}{2}$	0·0630	229K	0·18375	666K	0·525	1·90M
0·050	500	250	125	0·1260	114K	0·3675	333K	1·050	951K
0·075	750	375	$187\frac{1}{2}$	0·1890	76K	0·5512	222K	1·575	634K
0·100	1000	500	250	0·2520	57K	0·7350	166K	2·100	475K

Because change in gauge resistance produces a non-linear output which differs for tensile or compressive strains, it is necessary when measuring levels of strain from 1 to 10%, either to use two calibration resistors or make a correction for the single resistor. The equations in Table 10 (based on reference 18) enable calibration resistance values to be calculated for tensile or compressive strains for the bridge configurations shown.

In those situations where it is not possible to shunt the strain gauge locally, thereby necessitating remote calibration, certain errors can be introduced by neglecting the effect of the lead resistance in series with the shunt resistance. This important aspect is covered in reference 19.

When a shunt resistance is employed to calibrate a strain measurement system used to determine stress, then the following calibration path is assumed:

$$R_{sh} \rightarrow \varepsilon_s \rightarrow \varepsilon_m \rightarrow E \rightarrow \sigma$$

The shunt resistance (R_{sh}) produces a simulated strain (ε_s), which is assumed to be the same as the measured strain (ε_m), which is in turn assumed to be the same as that resulting from the true surface strain. The indicated strain (assumed to be the same as ε_m), is dependent upon the Wheatstone bridge input voltage, the gauge factor and the output relationship of the bridge active arms. Using the material constants (e.g. E), the stress (σ) is calculated. In the case of force determination from strain measurement, the calibration path becomes:

$$R_{sh} \rightarrow \varepsilon_s \rightarrow \varepsilon_m \rightarrow E \rightarrow \sigma \rightarrow F$$

where the force (F) is derived from the calculated stress (σ) from a knowledge of the csa.

From the calibration path it is possible to make up a traceability chain (Fig. 14) from the gauge back to National Standards, for all those quantities and those factors which affect them in the calibration path. Considering Fig. 14, note that there are only two links of the chain unbroken and that these are quantities which can be directly related to one of the six base units of measurement. They are:

(1) the bridge input and output voltages, which are units of electrical potential, derived from the base unit of electrical current;
(2) the material dimensions and constants; the dimensions being area (a derived quantity) and the constants obtained from length (a base unit), and force (a derived quantity).

Thus, when a parameter is a base unit of measurement, or a derivative thereof, and is measured on equipment calibrated to National Standards,

TABLE 10
HIGH STRAIN LEVEL CALIBRATION RESISTOR CALCULATION

Outputs and equivalent strains for equal-arm bridges

Bridge configuration and unbalance mode	Output equation for strain	Equivalent strain
¼ Bridge Only one arm unbalanced. (any one arm). Used to measure axial strain, tension or compression, in a uniaxial stress field. 	$$\left|\frac{V_{out}}{V_{in}}\right| = \frac{K\varepsilon}{4 + 2K\varepsilon}$$ Where the sign in the denominator becomes negative for a compression strain; the output being greater for compression than for tension.	for tension strain: $$(1) + \varepsilon = \frac{R}{KR_{sh}}$$ for compression strain: $$(2) - \varepsilon = \frac{R}{K(R_{sh} + R)}$$
½ Bridge Any two opposite arms unbalanced. Used to measure axial strain, tension or compression, in a uniaxial stress field. Increasing the number of active arms improves the signal-to-noise ratio.	$$\left|\frac{V_{out}}{V_{in}}\right| = \frac{K\varepsilon}{2 + K\varepsilon}$$ The above note for a 1/4 bridge applies to this axial bridge also.	for tension strain: $$(3) + \varepsilon = \frac{R}{K(2R_{sh} + 1/2R)}$$ for compression strain: $$(4) - \varepsilon = \frac{R}{K(2R_{sh} + 3/2R)}$$

½ Bridge

Two adjacent arms unbalanced. either arms 1 and 2 or arms 3 and 4. Used to measure pure bending, one arm in tension and other in equal compression. May be used similarly to measure pure shear or pure torsion.

$$\left|\frac{V_{\text{out}}}{V_{\text{in}}}\right| = \frac{K\varepsilon}{2}$$

If the two adjacent active arms are 1 and 4 or arms 2 and 3, the above equation does not apply and the output is non-linear.

$$(5)^a \pm \varepsilon = \frac{R}{K(2R_{\text{sh}} + R)}$$

Applies for arms 1 and 2 or arms 3 and 4 only.

Full Bridge

All four arms unbalanced. One pair of opposite arms in tension and other pair in equal compression. Used to measure bending, shear, and torsion as for 1/2 bridge just above.

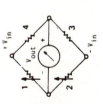

$$\left|\frac{V_{\text{out}}}{V_{\text{in}}}\right| = K\varepsilon$$

For all equations:
$V_{\text{out}}/V_{\text{in}} \times 10^3 =$ 'millivolts per volt'
$\varepsilon \times 10^6 =$ 'microstrains'

$$(6)^a \pm \varepsilon = \frac{R}{K(4R_{\text{sh}} + 2R)}$$

---

[a] For bending bridges, half of the arms are in compression and half in tension.
Table taken from *Exp. Mech.*, Oct. 1976, with permission.

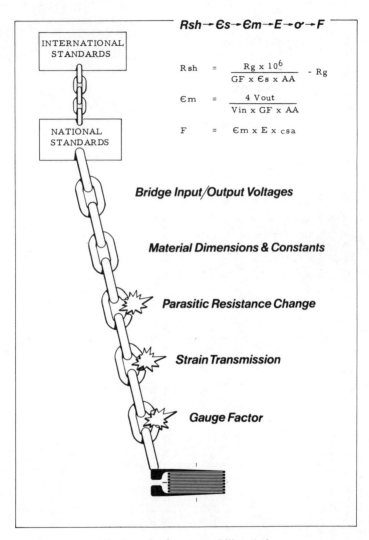

Fig. 14.   Broken traceability chain.

the accuracy of that measurement can be quantified, i.e. the parameter is an unbroken link in a traceability chain.

However, it will be observed from Fig 14 that for both stress and force determination there are three direct breaks in the traceability chain, the first two of which are associated with gauging practice.

(1) Parasitic resistance changes within the Wheatstone bridge circuit, which can be indistinguishable from the $\Delta R/R$ resulting from the surface strain $\Delta L/L$.

(2) Incomplete transmission of the surface strain through the adhesive layer and the gauge backing to the gauge foil.

(3) The break associated with the quoted manufacturer's gauge factor.

Gauge factor has been defined as 'The ratio of the proportional change in resistance of a strain gauge installation to the strain in the surface on which it is mounted, caused by uniaxial stress in the direction of the strain sensitive axis, all other variables remaining constant. Mathematically, gauge factor equals $(\Delta R/R)/(\Delta L/L)$, where L is the initial value of the active gauge length, R is the resistance of the strain gauge installation at L, and ΔR is the change in resistance R caused by a change, ΔL, in the active gauge length, L.'[20]

It will be noted that gauge factor is defined in two units: length, which is a base unit and electrical resistance which is a derived unit. This enables the gauge factor of the batch tested gauge to be determined, with traceability back to National Standards.

Each individual grid of a strain gauge installation has its own characteristics which are determined by the gauge manufacturer during the manufacturing process, and by the user during the installation (i.e. by gauging practice). These characteristics include:

(a) gauge factor, which is affected by temperature and strain level;
(b) transverse sensitivity;
(c) gauge resistance change when bonded to a specified material, due to temperature alone—often termed 'thermal output';
(d) stability of the bonded gauge system and the effect on zero drift of temperature, time at zero strain and time under strain (creep).

In the calculation of electrically simulated strain, the gauge factor is fundamental and, therefore, is a link in the traceability chain. This link is, however, broken by the nature of strain measurement using resistance gauges in that the gauge factor quoted by the manufacturer cannot be classed as truly traceable, since it was obtained by tests carried out on a representative number of gauges from a batch, the assumption (and hence the third broken link) being that the values obtained from the sample are valid for the whole batch. Thus, the manufacturer's quoted gauge characteristics are only stated on a statistical basis. The analogy of bullet testing amply illustrates the problem and shows the need for batch sampling techniques.

As already considered, the most important aspect of the gauge factor is the ability of the user to reproduce that quoted by the manufacturer with the gauge application techniques he employs in a particular 'on-site' condition. A batch sampling technique similar in principle to that used by the gauge manufacturer, but designed specifically to check the effects on a quoted gauge factor of the gauging techniques and adhesives employed for a specific job, is recommended.

There are a number of recognised methods of determining gauge factor. These include: (a) constant bending-moment beam; (b) uniform stress cantilever beam; (c) direct tension or compression.

The gauge manufacturer may employ any of the three methods, but requires a high degree of sophistication in equipment, techniques and environmental control to realise the necessary accuracy. The user may opt for a similar degree of sophistication if the costs are justifiable. However, this is rarely the case and other factors, including the lack of a suitable tightly controlled environment, could necessitate a much simpler rig which is less accurate than that used by the manufacturer. Irrespective of the rig type chosen, the data obtained from it should be fully traceable.

One suitable rig, proved in use for more than 15 years, is illustrated in Fig. 15. However, the rig proposed in reference 20 is of more recent design and will reduce the error sources associated with that of Fig. 15. Both of the above rigs are constant bending-moment types; from basic theory it can be shown that, neglecting second order errors

$$\text{longitudinal surface strain} = \frac{4dt}{a^2}$$

For convenience of calculation, the quantities t and a are selected such that $4t/a^2$ produces an integer N, so that $\Delta L/L = Nd$. For the rig illustrated in Fig. 15 $t = 0 \cdot 1280$ in and $a = 8 \cdot 000$ in therefore, $N = 8 \times 10^{-3}$ and $\Delta L/L = 8d$, where $d = $ bar deflection i.e. dial gauge reading (Fig. 16).

There is, however, one major problem with this type of four-point loaded beam in that it is often incorrectly assumed that the radius of curvature over the middle section is constant. The only constant is the bending-moment and this will only induce a constant surface strain in the middle section of a

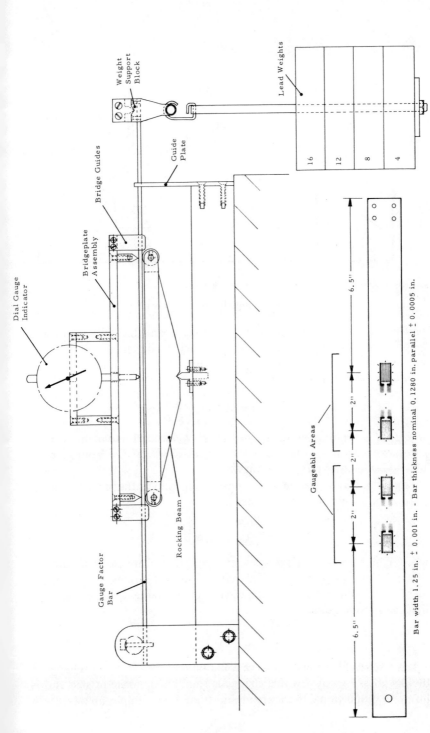

Weight Support Block

Lead Weights

16
12
8
4

Guide Plate

Bridge Guides

Bridgeplate Assembly

Dial Gauge Indicator

Rocking Beam

Gauge Factor Bar

Gaugeable Areas

6.5"

2" 2" 2" 2"

6.5"

Bar width 1.25 in. ± 0.001 in. - Bar thickness nominal 0.1280 in. parallel ± 0.0005 in.

FIG. 15. Gauge factor rig and bar.

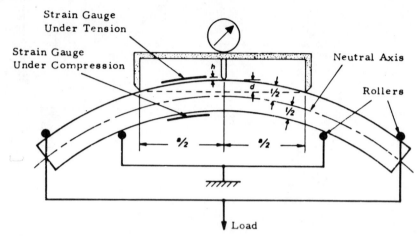

FIG. 16.   Gauge factor rig theory.

perfectly uniform beam. In cases where the beam thickness varies with length, the surface strain is inversely related to bar thickness, with the greatest surface strains produced in the thinnest sections, as shown below. Consider the beam equation

$$M = \frac{\sigma}{y} I = (E\varepsilon)\left(\frac{bt^3}{12}\right)\left(\frac{2}{t}\right)$$

where M = bending-moment, σ = stress; I = second moment of area, y = distance from the neutral axis, ε = surface strain, E = Young's modulus, b = breadth and t = thickness.

Since the bending-moment is constant, it follows that for the central portion of a four-point loaded beam

$$\varepsilon = \frac{6M}{bE} \times \frac{1}{t^2} = \frac{Q}{t^2}$$

where Q is a constant. (It is assumed that b and E are also constant.)

It follows from this equation that the surface strains at two points (denoted by suffices 1 and 2) are inversely related to the square of the beam thickness at the points, i.e.

$$\frac{\varepsilon_1}{\varepsilon_2} = \frac{t_2^2}{t_1^2}$$

It is also readily shown from the equation that the percentage change in the thickness t (provided this change is small) is opposite in sign and is inversely proportional to the change in strain. For example, an increase in bar thickness from 0·128 in to 0·129 in (i.e. a change of $+0·8\%$) will produce a decrease in strain of $1·6\%$ $((0·128)^2/(0·129)^2 = 0·984)$. With tolerances of $\pm 0·001$ in on a 0·128 in depth, changes of $\pm 1·6\%$ in the surface strain must be expected.[7]

Another consideration when using a four-point loaded beam is the usable working length of the beam. It has been suggested that only the centre two-thirds of the span between the inner rollers can be used. However, as reference 21 shows, as long as the entire gauge backing lies within the centre of the roller (for a tensile strain) or the gauge backing does not touch the roller (for a compressive strain), then the total length between the inner rollers can be used (Fig. 15). The usable working length may, however, be restricted by the location of the knife edges on the dial gauge bridge plate assembly.

The user's verification of the manufacturer's gauge factor is carried out by laying a gauge (for which the quoted gauge factor can be statistically linked to a traceably obtained batch gauge factor) on a reusable, accurately machined homogeneous gauge factor bar (Fig. 15). (Ideally, the gauge factor bar should be of the same material as that to which the gauges are bonded on site.)

The most representative results are obtained if the 'high quality' gauge is laid by the actual operator when the installation conditions and/or adhesive are at their worst from an ideal gauging point of view, e.g. at the end of a gauging exercise near the end of the usable life of the adhesive. It is essential that chemical surface preparation, gauge clamping arrangements, the adhesive and its curing cycle are similar for the test strain gauge to those used for the 'on-site' gauge installation, but excessive surface abrasion of the accurately machined bar should be avoided. The gauge factor bar is placed in the constant bending-moment gauge factor rig, and readings obtained for beam deflection and gauge resistance change in response to incremental loading. The dial gauge and the resistance measurement system used, must, of course, be 'in calibration' and traceable to National Standards. A basic resistance measurement, rather than direct readout in engineering units is preferred, to establish a more direct line of traceability.

If the calculated gauge factor does not agree realistically with that quoted by the manufacturer, say within $\pm 1\%$, it is necessary to ascertain the cause of the discrepancy: any correction factors should be checked, in particular the bar thickness correction; another gauge should be laid on a different bar and, if the discrepancy is repeated, the gauge manufacturer should be consulted, stating the gauge and adhesive batch numbers and the material and techniques employed for bonding the gauge.

Within the limitations imposed by the nature of strain measurement, this secondary chain of traceability is considered to be the best possible method of ensuring that the effects of the inevitable breaks in the main chain are minimised, and that reproducibility of the statistically based manufacturer's gauge factor can be achieved in practice by the user. A point worth considering at this stage is that, although the standard 'push-test' provides a useful verification that the gauge is bonded, it can never be regarded as a substitute for a job related verification of the quoted gauge factor. Simply, it just does not demonstrate the user's ability to reproduce the manufacturer's quoted gauge factor value, which is essential if any degree of credibility is to be given to a strain measurement system.

Error Sources in Gauge Factor Measurement

There are a number of identifiable error sources associated with the user derived gauge factor value obtained with the rig shown in Fig. 15, some of which are quantifiable. The individual errors vary in magnitude from that which may be considered insignificant to, in one case only, a value which is

TABLE 11

GAUGE FACTOR RIG ERRORS

Parameter	Realistic error	Major effect on gauge factor 2·1
Resistance measurement resolution	± 0.0005 ohm	± 0.007 gauge factor random
Resistance measurement linearity	± 0.005 ohm	± 0.07 gauge factor random
Dial gauge resolution	± 0.00005 in.	± 0.0044 gauge factor random
Dial gauge linearity	± 0.0005 in.[a]	± 0.012 gauge factor random
Micrometer reading resolution	± 0.0005 in.	± 0.008 gauge factor random
Micrometer reading accuracy	± 0.0005 in.[b]	± 0.016 gauge factor absolute
Bar width tolerance	± 0.0005 in.	± 0.001 gauge factor absolute
Bridge fulcrum distance, a	± 0.002 in.	± 0.001 gauge factor absolute
Gauge backing thickness tolerance	± 0.0003 in.	absolute ± 0.009 gauge factor
Adhesive thickness tolerance	Determined by user	Absolute
Temperature	Determined by user	Random and absolute

[a] Based on BS 907.
[b] BS 870.

greater than the 'typical' tolerance on the gauge factor quoted by the manufacturer. Absolute errors affect the accuracy of the user derived gauge factor and are indicated by differences between the manufacturer's quoted gauge factor and the user derived gauge factor. Random errors produce a spread in the incremental gauge factor values, which is evident from deviations of the individual values from the mean value.

Some of the errors are inherent in the rig itself, owing to machining, fitting tolerances, friction etc., whilst others are determined by the machining tolerances on the test bars and the displacement and resistance measurements. It is essential to ensure meticulous operation of the rig and that the bar is free to bend without touching the bar guides. The calculated gauge factor is not dependent upon the absolute value of the individual weights as they are only the means of producing the constant bending-moments.

In considering the effect of the various error sources it is necessary to assign an error value to each parameter. This is only possible if that parameter is in either a base unit or a derived unit. Table 11 lists the main error sources, with realistic values of error and the types of error they constitute, i.e. absolute or random. The Table is based upon three equations related to the rig:

(1)   Longitudinal surface strain at the dial gauge

$$\Delta L/L_1 = \frac{4t \, (\text{dial gauge reading})}{a^2}$$

(2)   Longitudinal surface strain at the strain gauge

$$\Delta L/L_2 = \frac{(t_1)^2 \, \Delta L/L_1}{(t_2)^2}$$

(3)   Gauge factor calculation

$$K = \frac{\Delta R/R}{\Delta L/L_2}$$

A calibration check-rig based on a constant rectangular section cantilever test beam is often used in strain measurement. This type of rig differs fundamentally from the gauge factor rig already discussed, in that it is not generally used to provide a traceable verification of the user's ability to reproduce the manufacturer's gauge factor (thereby establishing a secondary chain of traceability, necessary whenever calibration of a strain measurement system is by electrical simulation), but is used as a very useful

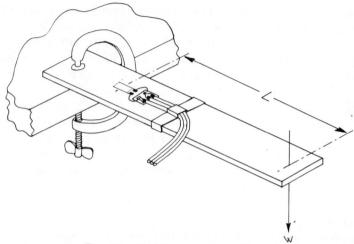

FIG. 17.   Cantilever test beam rig.

(and considerably cheaper) method of demonstrating the user's ability to achieve a predetermined strain value from a laid gauge. A typical cantilever beam is illustrated in Fig. 17 and the basic theory for calculation of strain is described as follows. Given that

$$\sigma = E\varepsilon = \frac{Mc}{I}$$

Therefore

$$\varepsilon = \frac{Mc}{EI} = \frac{WLh/2}{Ebh^3/12} = \frac{6WL}{Ebh^2}$$

where σ = stress (psi), ε = strain (in/in), E = Young's modulus, M = applied moment = WL, c = distance of surface above neutral axis = $h/2$ in, I = moment of inertia of rectangular beam = $bh^3/12$ in,[4] W = weight (lbs), L = length (in), h = beam thickness (in) and b = beam width (in).
Example: Aluminium beam (4 lbs weight) $5\cdot125$ in $\times\ 0\cdot1250$ in $\times\ 0\cdot75$ in. Therefore

$$\varepsilon = \frac{6 \times 4 \times 5\cdot125}{10\cdot6 \times 10^6 \times 0\cdot75 \times (0\cdot1250)^2} = 990\ \mu\,\text{in/in}$$

Correction for gauge and adhesive thickness (t): assume $t = 0\cdot0016$ in.

$$\text{Indicated strain} = \frac{\varepsilon(c + t)}{c}$$

$$= \frac{990\ (0\cdot065 + 0\cdot0016)}{0\cdot065} = 1015\ \mu\,\text{in/in}$$

It will be noted that the value of strain obtained from the beam is given by

$$\varepsilon = \frac{6WL}{Ebh^2}$$

and is, therefore, dependent upon knowledge of the weight (W), distance (L), Young's modulus of the beam (E), and the beam width (b) and thickness (h) at the gauge position. Owing to the nature of the strain distribution along the beam, the gauge output will be the average of the strain gradient affecting the active gauge length. The measured gauge output should, therefore, be equal to that calculated at a distance L, from the centre of the active gauge length, provided that b and h are constant over the entire gauge position area.

ERROR EFFECTS, MAGNITUDES AND CORRECTIONS

Conventionally, an error is given a positive or negative sign according to whether its effect is to increase or decrease the indicated or measured value. The actual magnitude, and therefore the effect, of any resulting errors from the numerous potential error sources associated with strain measurement are specific to a particular system and situation. It is necessary to conduct, for each measuring system, an 'error source investigation' to determine which of the listed 63 error sources (Tables 1 to 4), and any others, are applicable and which of these will result in significant errors (in the light of the calibration method and the environmental and technical considerations) and, the identifiable errors for which corrections can be made. Each individual case must be analysed and the remaining system uncertainty that is quoted must be qualified by the limitations used in determining the quoted value. Uncertainty is usually expressed in bilateral terms (i.e. $\pm$) either in units of the measured quantity or as a proportion (i.e. % or ppm) of the measured quantity or some other specified value.

To generalise about possible magnitudes of error (within the restraints given above) it can be stated that the magnitude of error in strain measurements with foil strain gauges depends on a combination of two basic parameters: (1) environmental conditions at point of installation and the measuring conditions; and (2) the expertise of the gauge installer or stress analyst.

As discussed in Chapter 3, there is usually little that can be done about the installation or test environment. The gauge manufacturer has obtained the batch sample gauge factor in a carefully controlled laboratory environment where the gauge has been bonded by highly skilled technicians to a flat specimen, subjected to a uniform stress field, tested statically at a low level of strain, supplied with low excitation voltage, connected to the instrument with lead wires of negligible resistance, etc., and yet the best he will quote is typically $\pm 0.5\%$. It can be stated, in general, that the greater the number and extent of deviations from these 'ideal' conditions, the greater will be the potential error and uncertainty.[22]

The position is probably best illustrated by the triaxial graph (taken from reference 22) in Fig. 18. This shows the approximate upper and lower

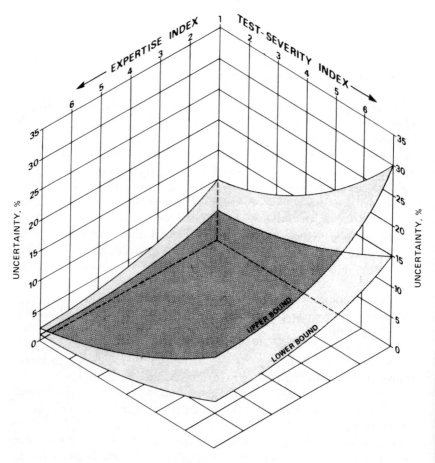

FIG. 18. Measurement uncertainty graph. Courtesy of Micro Measurements.

bounds of uncertainty to be expected as a function of test environment severity and the operator's ability. On the test severity scale, $1·0$ represents the 'ideal' laboratory conditions (similar to those used to obtain the gauge factor etc.), and this figure increases as the test environment becomes more severe. On the ability scale, $1·0$ applies to a complete novice with respect to practical strain gauge technology, and increasing numbers on this scale signify greater degrees of knowledge, experience and proficiency.

The graph indicates that under the best test conditions, with the most experienced practitioner and making corrections for all the known errors, the uncertainty can never be much less than $\pm 1\%$—$\pm 0·5\%$ uncertainty being contributed by the gauge factor tolerance alone. Under the same test conditions, an inexperienced operator is liable to produce an uncertainty in the range of ± 5–10%. The increased uncertainty may result from any one of a variety of causes, or a combination of such—incorrect gauge selection, for instance (too large a gauge length relative to the strain gradient, mismatch of the self-temperature-compensation characteristics, selection of isoelastic foil for static strain measurement etc.), lead wire resistance error, transverse sensitivity, mislocation or misorientation of the gauge, and others from the list of error sources (Tables 1–4). When test conditions are very severe, the uncertainty can reach ± 7–15% for the 'expert' and ± 15–30% (or more) for the novice.

While the uncertainty magnitudes employed in constructing Fig. 18 were assigned in a semi-arbitrary manner, there is good reason to believe that they are a reasonably accurate representation of the true state of affairs. Experience shows that at one extreme better than $\pm 1\%$ is very difficult to achieve, while at the other extreme (severe installation and test conditions, coupled with an inexperienced operator) it can be shown that errors of $\pm 30\%$ or more can readily result from failure to consider and control some of the error sources mentioned in this chapter.

To further reduce the total uncertainty in strain measurement it is possible to apply corrections for some errors. However, this is limited to those errors which obey known determinable laws, that are repeatable (i.e. category (2) errors) and for which the magnitude and sign of the error is known or can be determined.

It must always be borne in mind that the gauge data provided by the manufacturer, e.g. thermal output characteristic, was obtained using specified bonding techniques to a particular test material. It is necessary for the user to establish by experiment whether or not, with the techniques and materials he is using, there is substantial deviation from the quoted data. When computerised data reduction is being used (Chapter 4), the thermal

TABLE 12
CORRECTIONS TO STRAIN READINGS

Parameter effect	Error source	Method of correction	
Gauge factor	Use of common nominal K for rosette gauges	Use individual grid K quoted by manufacturer	K = Quoted gauge factor K' = New gauge factor R_g = Gauge resistance in ohms R_L = Resistance of lead in ohms two-wire—both leads three-wire—one lead only
	Change in K with temperature	Use manufacturer's quoted data	
	K desensitisation due to lead effects	Apply correction using: $K' = K\dfrac{R_g}{R_g + R_L}$	
	K desensitisation due to bridge balance components	Apply correction using: $K' = K\left[\dfrac{4R_b + R_g}{4R_b + 2R_g}\right]$	K = Quoted gauge factor K' = New gauge factor R_g = Gauge resistance in ohms R_b = Shunt balancing resistance in ohms
Temperature induced bridge output changes	Temperature effects on bridge balance components	Apply correction using: $S = \dfrac{\dfrac{(R_g + R_b)}{(R_g + R_b)} \times T_c \times (T_2 - T_1)}{4R_g K}$	T_c = Temperature coefficient of R_b $T_2 - T_1$ = Temperature difference of R_b S = Zero shift in microstrain
	Temperature effects on leads within the bridge circuit	Apply correction using: $S = \dfrac{R_L \times T_c}{\Delta R_g}$	S = Error in microstrain R_L = Resistance of uncompensated lead T_c = Temperature coefficient of conductor (Cu 0·004) ΔR_g = Change in gauge resistance for 1$\mu\varepsilon$ K of 2·0 120Ω 0·00024Ω 350Ω 0·0007Ω
	Change in gauge resistance due to 'thermal output'	Use manufacturer's quoted data or obtain data from laboratory experiments obtained in similar conditions as measurements. See WSM TN-128-2	
Non-temperature induced bridge output changes	Insulation breakdown after initial bridge balance	Apply correction using: $-K\varepsilon = \dfrac{\Delta R_l}{R_l}\left[\dfrac{R_g}{1 - \dfrac{\Delta R_l}{R_l}}\right]\dfrac{1}{R_l}$	K = gauge factor R_l = Insulation at balance ΔR_l = Change in insulation after balance
	Transverse sensitivity	Use manufacturer's data. See also WSM TN-137	
	Non-linearity	Use manufacturer's data. See also WSM TN-139	
	Gauge misalignment	Use manufacturer's data. See also WSM TN-138 and reference 13	
	Environmental effects	Use data obtained from laboratory experiments in similar conditions as measurements and on exercise technique	
Calibration	Non-linear effect of shunting	Use data in reference 18	
	Effect of lead resistance	Use data in reference 19	

output characteristics can also be supplied by the manufacturer in the form of a least-squares polynomial equation, in °F or °C.

Further corrections for the effects of temperature may be necessary, depending upon the degree of accuracy required, as the gauge factor also varies with temperature. This temperature-dependent aspect of gauge factor is difficult for the user to determine, so that it is usually necessary to use that quoted by the gauge manufacturer. A more detailed coverage of this subject is found in reference 23.

Corrections are usually made by either modifying, for various effects, the manufacturer's quoted gauge factor or by manipulation of the strain readings prior to the determination of stress values. Table 12 gives some of the main corrections which can be made to strain readings to reduce the effects of errors inherent in data obtained from foil gauges. The use of these corrections will obviously reduce the overall uncertainty of the measurements obtained.

ERRORS IN INSTRUMENTATION

The accuracy obtained from any measurement device is dependent to a large extent on the instrumentation system used with it. Indeed, the precision of any measurement can be no better than that of the instrument used for observation or recording of the output. The foil strain gauge is no exception to this, as is evident from the error sources cited in Table 3. However, it is worth noting that choice and use of instrumentation by the user is also a major source of potential error.

One essential point is that test measurements must be made with the same instrumentation, used in the same manner as for the calibration. Good examples of failure in this respect are using different galvanometers and changing bridge excitation modes. In the first case, the practicalities of calibration with a known force often mean transporting a gauged structure, with the associated instrumentation, from a laboratory to a test facility building and then to the test site, and entail removal of the ultra-violet galvanometers from the recorder block for safe transportation. If different galvanometers are inadvertently replaced in the recorder block for the test, then errors of up to $\pm 10\%$ can be experienced owing to individual spreads. The use of a simulated 'check calibration' will reduce this error as it enables the system sensitivity to be readjusted to that of the physical calibration. In the second case, a simulated calibration could be made using a constant voltage (CV) bridge excitation, but test data must be obtained with a constant current (CC) source. The manufacturer's gauge characteristics are usually obtained with a CV source and may alter if the gauge is energised with a CC source; the Wheatstone bridge output also is different for CV and CC sources, not being related to gauge factor when a CC source is used.[13]

It is, of course, impossible within the limitations of this chapter to identify all possible user induced errors, resulting from the misuse of instrumentation or inherent in the equipment itself. Suffice it to say that the examples quoted illustrate the problem and with some forethought and common sense application of measurement principles, the errors resulting from the instrumentation will contribute only minimum uncertainty to the overall measurement accuracy.

CONCLUSIONS

Errors in strain measurements made with metal foil gauges are inevitable owing to the nature of the materials and techniques employed. However, the magnitude and sign of these errors do, to a large extent, depend upon the choice of materials, the techniques with which they are used and the skill, knowledge and integrity of the operator installing the measuring system. It is essential that the operator is aware of potential error sources and the magnitudes of any possible resulting errors in the measurement system and also the methods available for eliminating, compensating, correcting or minimising these errors.

The method of calibration is decisive in determining whether or not the data obtained has a high or low level of uncertainty attached to it. A final example may help clarify the pitfalls of calibration by calculation. Four seemingly identical tension link type transducers, weighing 1300 N ($2\frac{1}{2}$ cwt), were calibrated by electrical simulation, and then for comparison (at considerable time and expense) against a known force. The results, illustrated in Fig. 19, show that in each case the simulated calibration was in error by -10%, -5%, -2% and $+1\%$ as a common csa value was used in the calculation. Had the check calibration with a known force not been carried out, the user could have been unaware of these errors, which might easily have been even greater.

The manufacturer's quoted gauge parameters, in particular gauge factor and thermal output, have been obtained using specified techniques and materials and are, therefore, only valid for installations which comply with the manufacturer's test conditions. Furthermore, the need exists for the user to demonstrate, for each different installation situation, his ability to

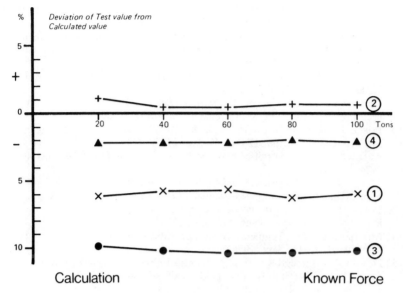

FIG. 19. Calibration deviation.

reproduce, within realistic limits, the quoted gauge factor of a gauge from a reputable manufacturer. The job related gauge factor bar fulfills that requirement when it is not possible to calibrate the measurement system directly against a known force induced strain. Although the bar can reduce strain transmission uncertainties, the technique does not remove the need for careful consideration of other error sources and corrections where applicable.

REFERENCES

1. DIETRICH, C. F., *Uncertainty, calibration and probability*, Adam Hilger, London, 1973.
2. CAMPION, P. J., *et al.*, *A code of practice for the detailed statement of accuracy*, HMSO London, 1973.
3. British Standards Institution, *British standard 5233 glossary of terms used in metrology*, London, 1975.
4. USSR State Committee on Standards GOST 8.0009-72, *Metrological characteristics of measuring instruments subject to standardization*, (in Russian), 1972, Moscow; English translation in NELTT 2417, NEL East Kilbride, 1972.
5. International Organization of Legal Metrology PD 1971, OIML, *Vocabulary of legal metrology*, Paris, 1969.

6. HAYWOOD, A. T. J., The state of the art of accuracy assessment: total confusion, *Control and Instrumentation*, Oct. 1977.
7. POPLE, J. and WRIGHT, T. D., The traceability of strain measurements using metal foil gauges, *BSSM/Inst. Prod. Eng. Conf.*, 1–5 Sept. 1980, University of Aston, Birmingham, England.
8. POPLE, J., DIY strain gauge transducers (Parts 1 and 2), *Strain* **16** (nos. 1 and 2), Jan. and April 1980.
9. *Optimising Strain Gauge Excitation Levels*, 1977, TN-127-3, Micro Measurements Division, Measurements Group Inc., Raleigh, USA.
10. POPLE, J., Increasing the output voltage of a Wheatstone bridge having one or two active arms, *Strain*, **12** (No. 1), Jan. 1976.
11. ARLOWE, H. D., A new pulsed power strain gage system, *ISA Trans.*, **13** (No. 3), 1974.
12. *Errors due to Wheatstone Bridge Non-linearity*, 1977, TN-139-2, Micro Measurements Division, Measurements Group Inc., Raleigh, USA.
13. POPLE, J., *BSSM strain measurement reference book*, BSSM 281, Heaton Road, Newcastle upon Tyne, England, 1980.
14. MORDAN, G. C., Temperature compensation methods for strain gauge transducers, *CME*, Sept. 1976.
15. DORSEY, J., Homegrown strain gage transducers, *Exp. Mech.*, July 1977.
16. STEIN, P. K., Traceability of the golden calf, *Measurements and Data*, **2**, Part 10, July–Aug. 1968.
17. BEHR, R. D., Why calculate simulated strain?, *Strain*, Nov. 1962.
18. TROKE, R. W., Improving strain measurement accuracy when using shunt calibrations, *Exp. Mech.*, Oct. 1976.
19. PERINO, P. R., The effect of transmission line resistance in the shunt calibration of bridge transducers, *Statham Labs Instr. Notes*, No. 36, Nov. 1959.
20. British Standards Institution, Methods of calibration of bonded electric-resistance strain gauges, *British Standards Institution DD6*, London, 1972.
21. COUTTS, J. A. and GREEN, K. P., Determination of the working length of a strain gauge calibration beam, *Strain*, **7** (No. 4), Oct. 1971.
22. PERRY, C. C., Uncertainties in strain gage measurements, *VRE Tech. Educ. Newsletter No. 26*, June 1979, Micro Measurements Division, Measurement Group Inc., Raleigh, USA.
23. *Strain Gauge Temperature Effects*, 1976, TN-128-2, Micro Measurements Division, Measurements Group Inc., Raleigh, USA.

PART II

Other Strain Gauges

Chapter 6

SEMICONDUCTOR STRAIN GAUGES

M. A. BAKER

Maywood Instruments Ltd, Basingstoke, UK

INTRODUCTION

In the 1950s experiments in a number of laboratories in the United States were conducted into the piezo resistance effect in semiconductors and confirmed that this effect could be much larger in semiconductors than conductors. Shortly afterward, a number of government laboratories and commercial enterprises undertook the development of strain gauges utilising this effect. Since then continuous development has led to the availability of a large number of satisfactory devices with excellent performance.

It is important to distinguish between piezo *electric* and piezo *resistive* phenomenon.

In piezo electric materials such as crystalline quartz, a change in the electronic charge (Q) across the faces of the crystal occurs when mechanically stressed. This effect is a dynamic phenomenon and the charge generally leaks away under static or quasi-static conditions unless buffered by very high impedance following circuitry. Thus these crystals are self-generating and their properties are used to good effect in a number of accelerometer designs.

The piezo resistive effect is defined as the change in resistance of a material due to an applied stress and this term is used commonly in connection with semiconducting materials. The effect was first reported by Lord Kelvin in 1856 in connection with the change in resistance of conductors when subjected to mechanical strain, a fact which has subsequently been exploited in the development of conventional metal wire and foil resistance strain gauges.

267

The piezo resistance effect is much larger in certain types of semi-conductors than conductors and leads to the production of strain gauges with gauge factors between 50 and 200, whereas most metal gauges have gauge factors no greater than 4·5 and more commonly around 2.

The resistivity of a semiconductor is inversely proportional to the product of the electronic charge, the number of charge carriers and their average mobility. The effect of applied stress is to change both the number and the average mobility of the charge carriers. The magnitude and sign of the charge depends upon the type of material, dopant type and level, carrier concentration and crystallographic orientation.

It can be shown that, for a metal conductor, the change in resistance due to strain induced dimensional changes is:

$$\frac{\Delta R}{R} = \frac{\Delta l}{l}(1 + 2v) \tag{1}$$

and the gauge factor, K

$$K = \frac{\Delta R/R}{\varepsilon} = (1 + 2v) \tag{2}$$

where ΔR = change in resistance, R = unstrained resistance, Δl = change in original length, l = original length, v = Poisson's ratio and $\varepsilon = \Delta l/l$ = strain. For most metals, the Poisson's ratio, $v \approx 0\cdot3$ and therefore $K = 1\cdot6$.

Since most metal strain gauges have an observed gauge factor between 2 and 4·5, a change in resistivity with applied stress must be postulated to account for the difference. It is this factor which predominates in some semiconducting materials.

For strain gauges made from a semiconductor material, the gauge factor is given by the expression

$$K = 1 + 2v + \pi_l E \tag{3}$$

where π_l = longitudinal piezo resistive coefficient (m^2/N) and E = Young's modulus of elasticity (N/m^2). It is self evident that the gauge factor is dependent not only on resistance change due to dimensional changes $(1 + 2v)$ but also due to the change in resistivity with strain $(\pi_l E)$. The characteristic π_l is determined only by the type and physical characteristics of the semiconductor material. By choosing the correct crystallographic orientation and dopant type, both positive and negative gauge factors may be obtained.

Silicon is now almost universally used for the manufacture of semi-conductor strain gauges although in a few specialist cases germanium may be used. For P-type silicon, gauge factors vary between $+100$ and $+175$, whilst N-type material produces gauges with gauge factors between -100 and -140.

MANUFACTURE

As previously mentioned, silicon is invariably used as the basic raw material for the manufacture of semiconductor strain gauges. High purity, single crystal silicon is required and Czochralski pulled crystals are normally used for gauge manufacture as they have greater crystal perfection than zone refined crystals. The electrical properties of silicon are determined by the type and density of additional elements added by 'doping' the parent material.

In pure silicon the carrier mobility is relatively low and therefore the resistivity of such an 'intrinsic' semiconductor is high. By adding, or doping, minute quantities of atoms from the third or fifth group in the Periodic Table to pure silicon the resistivity decreases dramatically by a factor of about 20. Two types of silicon may be produced by this process and are known as N- and P-type silicon.

(i)   *N-type*. The diffusion of phosphorus atoms leads to conduction by negative electrons and is known as N-conduction. Materials of this type exhibit a negative gauge factor.

(ii) *P-type*. Diffusion of boron atoms leads to conduction by positive 'holes', or vacancies, known as P-type conduction. Materials of this type exhibit a positive gauge factor.

Generally, the temperature coefficients of both gauge factor and resistance, together with linearity, are all improved with higher doping levels.

The silicon material is first of all produced as a continuous crystal approximately 50 mm in diameter. Slices are then cut from this ingot along a precisely defined crystallographic axis, generally (111) for P-type silicon, and (110) for N-type according to the Miller index. Invariably a flat is cut on one side of the crystal to assist in locating the correct crystallographic axis.

Diffusion of P- or N-type impurities is conducted by heating the slices of silicon in a radio frequency tube furnace to a temperature in excess of 1000 °C. An inert carrier gas containing minute quantities of the desired impurity atoms is then passed through the tube furnace and therefore around the silicon slices. The final doping level is determined by the dopant concentration and flow rate of the carrier gas together with the total elapsed time for the diffusion process. The silicon wafer is then progressively lapped and polished to a finished thickness of about 0·1 mm, and resistivity measurements made to determine and categorise the electrical characteristics of the processed material.

The pattern of gauges to be produced from the wafer is delineated by photolithographic masking and etching techniques, much in the manner of integrated circuit production.

In high quality gauges, a layer of silicon dioxide is grown over the surface of the slice with 'windows' over the areas where electrical connections are to be made. An additional diffusion process is then conducted ensuring that these areas are very heavily doped. These local, low resistivity areas, ensure that 'diode' effects at the junction of silicon and leadout wire are minimised. Reverse biasing the junction results in a resistance variation of less than 0·5 Ω in correctly processed material.

By further masking processes, connection pads are formed at each gauge connection point by evaporating a thin film of metal, usually aluminium or gold, on the surface. Normally, electrical connections to the gauges are made before separation from the wafer by welding 99·99 % pure gold leads to the deposited pads by ultrasonic or thermocompression ball bonding processes (Fig. 1).

Finally, separation of the gauges from the wafer is completed by a combination of delicate lapping and etching processes. Great care is needed at this stage as the ultimate strength of the gauge filament is largely determined by the presence of surface dislocations which act as stress raisers reducing both fatigue life and ultimate stress. Gauges manufactured by modern processes commonly exhibit a fatigue life of $>10^8$ cycles at $\pm 1000\,\mu\varepsilon$ and an elastic strain range $> 5000\,\mu\varepsilon$.

The thickness of a finished strain gauge is also crucial if faithful transmission of strain from specimen to gauge is to be accomplished. Modern transducer quality strain gauges have finished thicknesses of the order of 0·01 mm and may readily be wrapped around a 2 mm radius, without fracture, to illustrate their flexibility.

Semiconductor strain gauges thus far described are strips of silicon either straight or U-shaped (Fig. 2) and are devoid of any backing material such as is commonly associated with metal foil gauges. This form of gauge has the highest performance because of the lack of a plastic backing, but is, in general, a more difficult form for the inexperienced technician to use. This is particularly true if gauges have to be installed outside a laboratory environment. To cater for this contingency, semiconductor gauges are also

FIG. 1. Silicon slice showing gauges after metallisation. Courtesy of Micro Gauge Inc.

available fully encapsulated in an epoxy-glass or similar matrix, together with integral leadout wires (Fig. 3).

Silicon strain gauges are normally supplied in sets of four produced from the same wafer of material, with resistance matched to a close tolerance (typically $\pm 2\%$) at a fixed reference temperature, normally room ambient.

At first sight, this process would be thought to produce gauges evenly matched in resistance variations. However, due to finite variations in the evenness of diffusion across the wafer of original material, selection of gauges from a single wafer does not necessarily produce optimal resistance tracking, one gauge to another, over wide temperature excursions. At least one major manufacturer (Micro Gage Inc.) has largely overcome this limitation by the employment of a sophisticated computer matching technique. Perhaps as many as one thousand individual strain gauges may

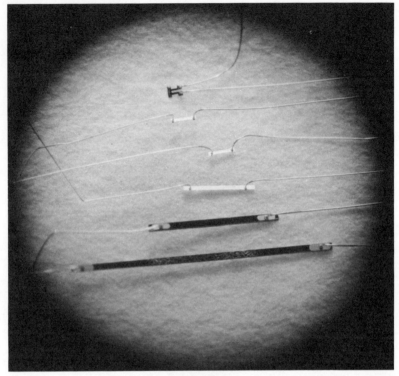

FIG. 2.   Typical semiconductor strain gauges. Courtesy of Micro Gauge Inc.

be made from a single silicon wafer. In the computer matching process, a
batch of gauges are loaded into special test jigs which are, in turn, loaded
into a thermal chamber. The gauges are then carefully tested over a 110 °C
range and the resistance variations of each gauge data logged at three
different temperatures. This data is finally computer analysed and sorted
into 'best-fit' sets of four gauges. The selected sets of gauges exhibit virtually

Encapsulated   Gauge

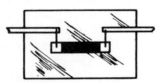

FIG. 3.   Typical encapsulated semiconductor strain gauge.

identical characteristics of resistance versus temperature and, in transducers utilising these gauges, the bridge 'zero' output is essentially compensated without recourse to secondary compensation techniques. This technique realises real matched sets of gauges matched to $\pm 1\%$ of nominal resistance at room ambient temperature and $\pm 0.2\%$ over a $110\,^\circ C$ temperature range.

SENSITIVITY

In an earlier section it was shown that the strain sensitivity of a semiconductor material was given by the expression

$$K = 1 + 2v + \pi_l E \tag{4}$$

where $\pi_l E$ is the most dominant term.

The high strain sensitivity of semiconductor strain gauges can, in essence, be reduced to anisotropic variations of the carrier mobility in silicon and germanium. In an electric field, a valence electron can be excited into the conduction band where it participates in the conduction of electric current. However, a relatively high excitation energy of about 1 eV is needed to promote conduction and the resistivity of these 'intrinsic' semiconductors is also relatively high. If pure silicon or germanium, both from the fourth group of the Periodic Table, are 'doped' with minute quantities of atoms from either the third or fifth group, the resistivity drops dramatically.

(a) *N-conduction.* Atoms of the fifth group produce four perfect bonds with the silicon atom with a fifth only weakly attached. An energy of only 0·05 eV is necessary to excite this valence electron into the conduction band. This process is known as N-conduction, the dopant element is normally phosphorus, and the resulting strain sensitivity or gauge factor is negative with applied strain.

(b) *P-conduction.* Similarly, if atoms of the third group, typically boron, are used as the dopant element there are three perfect bonds. An energy of only 0·08 eV will fill the vacancy or hole at the fourth bond. This process is known as P-conduction and leads to the production of semiconductor strain gauges with positive gauge factors.

The effect of doping with either P- or N-type impurities is to lower the resistivity as a result of the lower energies needed to initiate the conduction process. In general, the gauge factor, resistance, and temperature coefficient of both gauge factor and resistance all reduce with increased doping levels.

The other most important factor upon which conduction depends is the crystallographic orientation of the material from which the strain gauge is manufactured and it is mandatory that this be known and controlled during the manufacturing process. In silicon, now almost universally used for semiconductor strain gauges, P-type gauges are made from material orientated in the (111)* direction whilst N-type gauges are from (100)* orientation.

LINEARITY

In contrast to wire and foil gauges, semiconductor strain gauges can be distinctly non-linear. The relative resistance change at constant temperature can be written in the form of a polynomial:

$$\frac{\Delta R}{R} = C_1 \varepsilon^1 + C_2 \varepsilon^2 + C_3 \varepsilon^3 + \cdots \tag{5}$$

where $C_1, C_2, C_3 \ldots$, are constants and ε is the strain. For practical strain levels the strain cubed term and higher orders can be ignored as their contribution to the overall equation is negligible.

Typical strain sensitivity curves for both P- and N-type semiconductor strain gauges are shown in Figs 4 and 5.

Non-linearity can be appreciably improved by the selection of heavily doped material of lower resistivity and approaches that, over normally experienced strain ranges, of metal strain gauges.

When semiconductor gauges are used in either half active or fully active Wheatstone bridge circuits, appreciable, though not complete, compensation of gauge non-linearity is achieved through cancellation of the non-linear terms in the transfer function. Thus, in typical transducer applications where semiconductor strain gauges are used, non-linearities of less than 0.05% full scale are readily achievable, given good mechanical design. It is in $\frac{1}{4}$ bridge circuits, commonly found in stress analysis applications, where most problems associated with non-linearity are likely to be encountered.

In this specific application however, the selection of semiconductor

* Miller Index.

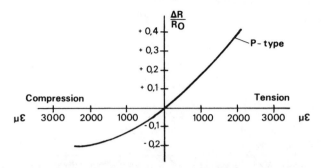

FIG. 4.   Resistance change versus strain for a P-type semiconductor.

strain gauges will have been predicated on the assumption that low strain levels are present, below the levels that can be accurately monitored using conventional gauges. By reference to Fig. 6 it may be readily discerned that the non-linear part of the strain sensitivity curve lies outside the range of ± 1000 microstrain and so, in general, non-linearity should not be a problem in most practical applications.

Caution should be exercised when interpreting strain sensitivity curves as the installed strain gauge may already be under a prestrain due to the bonding method employed. This has the effect of moving the unstrained reference point for the gauge along the curve, the amount by which this occurs can be readily determined by comparing the bonded to the unbonded gauge resistance. The difference in resistance can be used to determine the static prestrain present in the bonded strain gauge. This phenomenon is often, erroneously, thought to be due to cement shrinkage during the curing cycle. The true cause is to be found in the differential thermal coefficient of linear expansion of the gauge and the substrate to

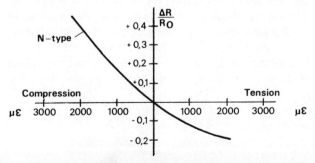

FIG. 5.   Resistance change versus strain for a N-type semiconductor.

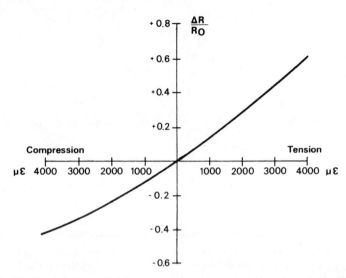

FIG. 6. Strain sensitivity curve for typical P-type semiconductors.

which it is bonded. Silicon has a lower coefficient of expansion than most common structural materials and so, if a high temperature curing epoxy is used as the bonding agent, then at the curing temperature a differential strain will exist. On return to room temperature after curing, the gauge will be put under a compressive strain, reducing its resistance and shifting the datum point for the gauge towards the compression half of the strain sensitivity curve.

In some applications, particularly in transducer design, it may be necessary to employ four gauges in a Wheatstone bridge where two of the four gauges are subject to a fraction of the strain applied to the other gauges. Typical examples are to be found in load cells employing the compressive column principle or diaphragm type pressure transducers. It is not always realised that the electrical output of a partially active Wheatstone bridge is intrinsically non-linear and the effect is more noticeable in semiconductor bridges because of the much higher strain sensitivity.

The output sensitivity of the Wheatstone bridge configuration shown in Fig. 7 is given by

$$\frac{e_0}{v} = \frac{\Delta R(1 + A)}{2[R + \Delta R/2(1 - A)]} \tag{6}$$

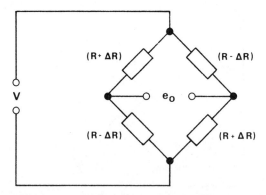

FIG. 7. Wheatstone bridge circuit with partially active gauges.

where A is the degree of activity of the partially active gauge. Any electrical non-linearity may be calculated from this expression although, for all but the most stringent applications, the effect will probably be negligible except where large asymmetries exist between gauges.

THERMAL CHARACTERISTICS

There are two basic thermal errors which apply to semiconductor strain gauges, namely the thermal coefficient of resistance (TCR) and the thermal coefficient of gauge factor (TCGF). To properly utilise semiconductor strain gauges in either stress analysis or transducer applications it is necessary to recognise and compensate for these effects. Compensation can generally be realised with simple resistive compensation circuits or other circuit artifacts and descriptions of a number of techniques follow. It is important to realise that although these techniques differ in concept from those employed in foil gauge practice they are nevertheless equally simple in implementation. The most common cause of difficulty in thermally compensating semiconductor strain gauges is the mistaken employment of foil gauge techniques which are only marginally effective.

Thermal Coefficient of Resistance (TCR)

Figure 8 illustrates a symmetrical Wheatstone bridge of four P-type semiconductor gauges. When an excitation voltage is applied to the input terminals, the no-load output voltage will be zero only if the bridge is exactly balanced:

$$R_2 R_4 = R_1 R_3 \qquad (7)$$

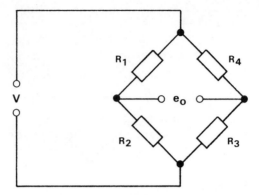

FIG. 8.   Symmetrical Wheatstone bridge circuit.

Furthermore, the no-load output will shift with temperature unless the effective TCR of each arm is identical, or at least obeys the condition that

$$\left(1 + \frac{\Delta R_2}{R_2}\right)\left(1 + \frac{\Delta R_4}{R_4}\right) = \left(1 + \frac{\Delta R_1}{R_1}\right)\left(1 + \frac{\Delta R_3}{R_3}\right) \tag{8}$$

over the operating temperature range.

Practically, it is impossible to bond a set of four gauges to a substrate and ensure that each gauge resistance and bonded TCR are identical. Therefore, any practical bridge will display some zero output and thermal zero shift, and these errors being essentially random in nature, can be of either sign. Compensation can be provided by leaving one corner of the bridge open (Fig. 9).

Equation (8) indicates that an apparent strain or no-load output is fundamentally due to a differential temperature coefficient of resistance across the bridge. By connecting a fixed resistor of low thermal coefficient across the bridge arm with the higher temperature coefficient will effectively reduce the overall temperature coefficient of that arm. By careful selection of resistors the TCR of the bridge arms can be matched, thereby compensating for thermal zero shift. Simultaneously, the no-load output can be reduced to zero by fitting a series resistor across the open node of the bridge as illustrated in Fig. 9.

In practice, the compensation may be realised either by theoretical calculation based upon measurement of actual thermal performance, or by iterative substitution of resistors. Difficulties may be encountered in predetermining the thermal characteristics of the bonded gauges therefore the second method is usually adopted. At first sight this may seem a tedious

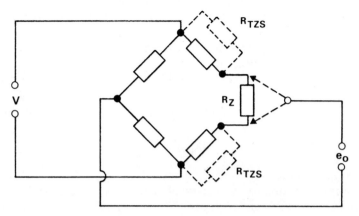

FIG. 9. Thermal zero shift compensation circuit.

procedure but in fact it is quite straightforward. The procedure is illustrated in the Appendix.

From the practical viewpoint, apparent strain bridge voltage outputs should be examined with care before deciding that the bridge has an unacceptably high level of apparent strain with temperature. For example, a bridge with an excitation voltage of 5 volts and four active gauges with a gauge factor of $+100$ may give an output of 3 mV over a temperature change of 50 °C. This apparently high output by foil gauge standards equates to only $0.12\ \mu\varepsilon/°C$ and is often misinterpreted by users new to semiconductor technology.

In transducer applications utilising modern computer matched gauges, zero temperature coefficients of less than $\pm0.015\%/°C$ are readily achieved without compensation. Accuracies to $\pm0.001\%/°C$ are achievable by further compensation using techniques similar to those enumerated above.

Zero Shift—Single Gauge Circuits
In $\frac{1}{4}$ bridge circuits commonly used in stress analysis applications compensation using push–pull gauge pairs is not possible unless a dummy gauge can be sited isothermally with the measuring gauge, often not possible in practical installations. In this case several alternative means are possible.

Self-Compensation Using N-Type Gauges
Unlike metal strain gauges, the temperature coefficient of resistance of elemental silicon and germanium cannot be modified to compensate for the

expansion coefficient of common substrate materials; however, N-type gauges can offer similar characteristics. For an accurate appreciation of the thermal effects acting on semiconductor materials, the TCR and TCGF values must be considered along with the properties of the material to which the gauge is bonded. The thermal coefficients of semiconductor strain gauges are quoted in manufacturers literature and invariably apply to the *unbonded* gauge only. When bonded to another material the *bonded* TCR is given by

$$\alpha_b = \alpha + (C_m - C_s)K \qquad (9)$$

where α_b = bonded TCR, α = unbonded TCR, C_m = thermal expansion coefficient (substrate), C_s = thermal expansion coefficient (gauge), K = gauge factor.

Table 1 lists typical thermal properties for semiconductor gauges.

TABLE 1

	P-type			N-type			
K	α %/100°C	β %/100°C	K	α %/100°C	β %/100°C	C_0	
100	7·2	−10·8	−100	3·6	−21·6	6·1	
115	5·4	−14·4	−110	10·8	−23·4	12·2	
130	10·8	−18·0	−120	23·4	−27·0	22·0	
140	18·0	−19·8	−130	28·8	−32·4	24·7	
155	32·4	−23·4	−135	43·2	−36·0	34·6	

N-type gauges have a negative gauge factor. Thus the bonded TCR of eqn. (9) can be zero, or negative when the differential expansion term exceeds the intrinsic TCR thus

$$C_0 = C_s + \frac{\alpha}{|K|} \qquad (10)$$

Self-compensation is accomplished when $C_0 = C_m$. For the example, an N-type gauge of gauge factor -110 will be essentially self-compensated when bonded to mild steel. ($C_m = 11 - 12\,\text{ppm}/°C$; $C_0 = 12·2\,\text{ppm}/°C$).

Self-Compensation Using P–N Pairs of Gauges

Dual element pairs of gauges consisting of a P-type gauge and an N-type gauge are available from some manufacturers. These gauges in adjacent arms of a Wheatstone bridge can be chosen to be self-compensating for a specific expansion substrate material. The thermal zero output voltage will be zero when

$$C_m = C_s + \frac{\alpha_N - \alpha_P}{|G_N| + G_P} \qquad (11)$$

where the subscripts P and N refer to the associated gauge types.

Thermal Coefficient of Gauge Factor (TCGF)

The gauge factor, K, is primarily determined by the doping level and, furthermore, varies as a function of temperature. Figure 10 illustrates the effect of doping level on gauge factor, and temperature dependence of gauge factor. By examination it may be determined that both the gauge factor and the temperature coefficient of gauge factor are inversely proportional to the doping level. Most commercial silicon strain gauges use relatively heavy doping levels with gauge factors of between $+100$ and $+140$ and at these levels the TCGF is reasonable. The casual observer may ask why heavier doping levels of the order $0·001 \, \Omega$ cm are not used, as the TCGF is approximately zero. However, present diffusion processes are not capable of producing this impurity concentration uniformly on a production basis.

When considering the compensation of TCGF it is again necessary to take into account the interaction between the gauge and the substrate to which it is bonded. Most common structural materials have a temperature coefficient of Young's Modulus, sometimes known as the 'thermoelastic coefficient'. Thus:

$$\beta_b = \beta + \gamma \qquad (12)$$

where β is the unbonded TCGF, β_b is the bonded TCGF and γ is the thermoelastic coefficient.

Compensation of TCGF

Several methods are available for the compensation of TCGF errors.
Series resistance method. The first, simplest method for use with constant voltage circuits, involves the fitting of a series resistor, usually in two equal halves to maintain bridge symmetry, in the excitation leads to the bridge as shown in Fig. 11. This circuit relies for its operation on two facts fundamental to the properties of silicon strain gauges. Namely, the TCGF is *always* negative whilst the TCR is *always* positive. It will be readily realised that in order to maintain constant sensitivity, an increase in bridge

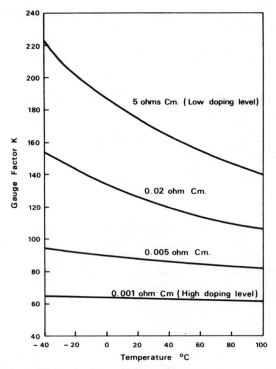

Fig. 10.   Effect of doping level on gauge factor and temperature dependence of
gauge factor.

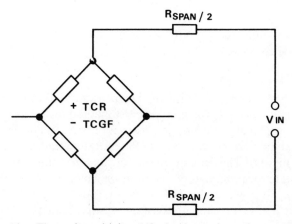

Fig. 11.   Thermal sensitivity shift circuit—series resistor method.

excitation with temperature is required. This may be accomplished, with constant voltage excitation, by fitting a low TCR resistor in series with the bridge. If the circuit is now considered as a voltage divider with the bridge treated as a single, positive TCR resistor, the bridge voltage will be caused to vary in proportion to the ratio of the two resistances. Thus, if the ratio of bridge resistance to series resistance is chosen correctly, the increase in bridge voltage due to the positive TCR of the bridge, may be made to offset the negative sensitivity coefficient.

The value of the series sensitivity compensating resistor may be derived from the following expression:

$$\frac{R_s}{R_G} = \frac{|\beta_b|}{\alpha_b - |\beta|} \tag{13}$$

where $\dfrac{R_s}{R_G} = \dfrac{\text{Sensitivity resistor}}{\text{Bridge resistance}}$

α_b = bridge temperature coefficient of resistance (eqn. (9)) identical to the individual gauge bonded TCR for a full bridge circuit; $\frac{1}{2}$ bridge circuits will have a TCR of 50% gauge TCR.

$|\beta_b|$ = the magnitude of the bonded TCGF (eqn. (12)).

An important advantage of both this method, and the preceding method for thermal zero shift compensation is that *thermally insensitive* resistors are used for compensation. Therefore, it follows that the siting of these resistors relative to the thermal environment of the strain gauge bridge is unimportant, ensuring freedom from thermal transient effects and warm-up drift.

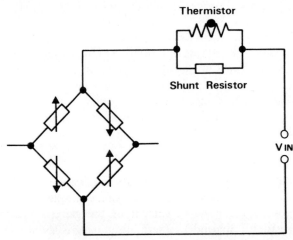

FIG. 12.   Thermal sensitivity shift compensation circuit—thermistor method.

Thermistor (NTC resistor) method. A second method, less wasteful on power, involves the use of a negative TC thermistor and shunt resistor as shown in Fig. 12. However, the thermistor has to be in the same thermal environment as the bridge.

Constant current method. Gauges can be selected with bonded TCR and TCGF of equal magnitude and opposite sign. If a bridge of such gauges is excited with a constant current supply then the bridge will be self-compensated. Adjustments to the compensation can be made by the addition of a shunt resistor across the bridge excitation leads as an analogue of the constant voltage scheme as described under the heading 'Series resistance method'.

OTHER PROPERTIES

Gauge Excitation
The maximum voltage that can be applied to a semiconductor strain gauge is determined by self-heating considerations and therefore dependent on the surface area of the gauge, the heat sink conditions and the gauge resistance. It is therefore difficult to make generalisations about optimum excitation levels, but for most transducer applications a power dissipation of about 10–20 mW per gauge will normally be found satisfactory.

Mechanical Properties
Comparison of metal foil gauges and silicon strain gauges reveals that significantly improved properties are present in certain areas.

Fatigue Life and Maximum Strain Range
Both the fatigue life and the maximum strain range are highly dependent on the micro surface finish of the silicon crystal. Gauges manufactured by modern etching processes commonly exhibit a fatigue life of $> 10^8$ fully reversed cycles at ± 1000 microstrain and minimum elastic strain range of 5000 microstrain. Bare gauges can be applied to radii as low as 3 mm without damage.

Hysteresis and Creep
Mono-crystalline silicon is a perfectly elastic material with no plastic region below 500 °C. Any hysteresis or creep in a gauge installation will normally be a function of the bonding cement or the gauged structure itself. Similarly, resistivity dependent thermal characteristics are also totally

repeatable and any thermal instability in a finished transducer element will be due either to defects in the transducer element or the bonding process.

Photoelectric Sensitivity

The effect of high ambient light levels on semiconductor materials is to marginally increase the number of charge carriers reducing the resistivity of the material. As would be expected, this effect is more marked in higher resistivity material and may amount to about 2–3 microstrain between darkness and normal daylight. Gauge installations should be protected from light for most accurate results.

SEMICONDUCTOR STRAIN GAUGE INSTALLATION

Silicon strain gauges are normally supplied in the form of straight or U-shaped thin strips of silicon with two welded gold leads attached (Fig. 2). Bare gauges ensure the most intimate contact of the gauge with the surface to be bonded and are recommended for transducer and other critical applications. Encapsulated gauges are also available from some manufacturers, with the silicon element sandwiched in an epoxy–glass, or similar matrix, together with more robust leadout wires. These are generally used for stress analysis work and, due to their construction and larger size, are not ideally suited to transducer applications. The bonding of encapsulated semiconductor gauges follows exactly the same procedure as that employed with metal foil gauges, details of which can be found elsewhere.

When bonding either encapsulated or bare semiconductor gauges fastidious attention should be given to surface preparation as the quality of the finished assembly is totally dependent upon correct surface preparation.

Surface preparation follows the same principles as those employed with foil gauges and again details will be found elsewhere. After the surface has been cleaned and prepared a layer of strain gauge adhesive is applied and cured as a precoat. This difference in procedure is necessary, as the bare semiconductor filament does not have an insulating backing to electrically isolate it from the bonded surface. Many commercially available strain gauge epoxies are suitable but it cannot be over emphasised that only epoxies specially formulated for strain gauge bonding should be used. Inferior results will accrue from using unsatisfactory adhesives. The next stage in the bonding process is to abrade the cured precoat surface which will generally have a glossy surface. A convenient way of achieving an ideal surface is to abrade the surface with pumice powder applied on the end of a cotton bud which has been moistened with neutraliser. The resulting matt surface is finally cleaned with conditioner and neutraliser in the normal way. This procedure will produce an ideal insulating surface about 0·01 mm thick ready for bonding the gauge. Bonding the strain gauge is carried out by coating the surface where the gauge is to be bonded and the gauge itself. The bare gauge is merely placed into position; no clamping is used as capillary action alone, between the gauge and precoated surface, is sufficient to provide thin glue lines even on overhead surfaces. With care and experience, glue line thicknesses of 0·01 mm are entirely feasible. The gauged assembly should now be cured following the manufacturers recommended cure schedules.

Although the above procedure may, at first reading, appear quite involved, in fact, experienced operators are able to bond semiconductor

FIG. 13.   Semiconductor strain gauge installation.

gauges more rapidly than foil gauges, mainly due to the absence of clamping procedures.

Wiring to Semiconductor Strain Gauges

Unencapsulated semiconductor strain gauges are invariably supplied with leads already welded to the gauge filament, the lead material usually being 99·99 % pure gold and about 0·05 mm in diameter. The normal method of connecting the gauge leads is to first of all bond a miniature strain gauge terminal tab near the gauge and then to solder the leads to this. Normal 60 % tin/40 % lead solder should not be used because the gold wire dissolves very rapidly in the lead constituent. The use of 95 % tin/5 % silver solder avoids this difficulty and is as easy to use as normal tin/lead solders.

Figures 13 and 14 show typical semiconductor strain gauge installations.

FIG. 14. Installation of four semiconductor strain gauges. Courtesy of Micro Gauge Inc.

OTHER BONDING METHODS

Although epoxy resins are by far the most widely used bonding agents several other advanced processes have become available to the transducer manufacturer. Discussion of these processes will be limited as they are outside the scope of the general strain gauge user due to the specialised equipment required.

As stated earlier single crystal silicon is a perfectly elastic material and any hysteresis in a strain gauged assembly can therefore be attributed to either the component itself or the bonding cement. In an effort to find a better bonding material, glass has been utilised to good effect in two differing processes.

Solder Glass Bonding

Relatively low melting point glasses of the type known as Pyroceram, manufactured by Corning Glass, have been successfully used to bond semiconductor gauges to metal substrates. Pyroceram is a vitreous, crystallising material which devitrifies during curing. Heating to a temperature above the softening point causes the glass to flow, and as the heating is continued, crystalline glass is formed. The process consists of four basic steps:

(i)   Finely powdered glass is mixed with a suspension vehicle.
(ii) The frit is placed on the metal component where the gauge is to be bonded.
(iii) The semiconductor gauges are placed in position and the assembly heated to evaporate the solvent suspension vehicle.
(iv) The structure is fired at 430 °C to effect the bond.

Although this method produces gauged assemblies with effectively zero hysteresis and high levels of repeatability, one major disadvantage exists. Most metals to which gauges are bonded have thermal expansion coefficients greater than silicon ($2.5 \times 10^{-6}/°C$) and when the bond is made at an elevated temperature, in this case 430 °C, the gauge will be subjected to a compressive prestrain when cooled to room temperature. This may amount to 3500 microstrain on common structural materials and therefore this method may be more successful on low expansion coefficient alloys. A secondary consideration is that all types of lead attachment may not be compatible with the high temperatures involved.

Electrostatic Bonding

A second approach makes use of a recently developed electrostatic bonding technique (US Patent 3 397 278) to effect a bond between silicon and glass. Developments of this technique can also effect a strong hermetic seal between glass and various metals. Bonding is accomplished by placing an ultra thin polished glass wafer between the gauge and the metal surface and simultaneously applying heat and a strong electrostatic bias across the assembly. For the process to be successful it is necessary that the glass wafer, strain gauge and substrate be chemically clean and polished optically flat. The fine surface finish ensures that the gap between the various assembled components is very small indeed and the electrostatic field produces immense forces of attraction. Under these influences the mating surfaces fuse together molecularly at a temperature lower than that required for the glass bonding process. This process produces superb results but is applicable only to specially manufactured components, precise process control is mandatory and special gauges with optically flat backs are required. At the moment this process is only used in certain specialist transducer applications and leads to the production of devices with zero hysteresis and effectively perfect thermal and mechanical repeatability together with excellent long term stability.

ADVANTAGES OF SEMICONDUCTOR STRAIN GAUGES IN TRANSDUCER APPLICATIONS

High sensitivity. A gauge factor of 50 to 70 times that of a metal foil gauge enables the design of transducers with high sensitivity. Alternatively, an enhanced dynamic response may result from the design of stiffer transducer elements.

Low hysteresis. Single crystal silicon has zero hysteresis, actual hysteresis is limited by bonding technique.

Wide resistance ranges. Gauges are available with resistances from 10 ohms to 10 000 ohms.

Gauge factor. Choice of gauge factor may be made between 50 and 175. The gauge factor may also be positive or negative depending upon whether P- or N-type silicon has been used to manufacture the gauge.

Small size. High resistances available in small sizes minimising power dissipation and self-heating effects, thus enabling miniature transducer

elements to be manufactured. Actual size may be as small as 0·5 mm long ×
0·1 mm wide.

High fatigue life. Fatigue life typically greater than 10^8 fully reversed
cycles at ±1000 microstrain.

APPENDIX

Thermal Zero Shift Compensation Method

(1) Wire the bridge according to Fig. 9.

(2) Connect a substitute resistance box across the open node of the
bridge.

(3) Switch on the excitation voltage and adjust the resistance box until
a zero output is obtained, note the resistance value and the arm of
the bridge in which it is connected.

(4) Place the gauged assembly into a thermal chamber and raise the
temperature.

(5) Note the new no-load output voltage and its polarity.

(6) Attach a second resistance box in shunt across the bridge arm with
the higher TCR as indicated by the polarity of the thermal zero
shift.

(7) Substitute a series of values of shunt resistor, at each substitution
adjusting the zero resistance until a zero output is obtained.

(8) When a tabulation of these results has been obtained reduce the
temperature of the assembly to the original as in step (3).

(9) Substitute the same pairs of shunt and series resistances as obtained
in step (7); only one pair of resistances will give a null output at
both high and low temperatures.

(10) Fixed resistors should now be fitted permanently to the bridge in
the appropriate positions. The resistors used should have a low
TCR, typically below 100 ppm. Metal glaze, metal film and thick
film cermet resistors are all suitable types.

Chapter 7

CAPACITANCE STRAIN GAUGES

E. Procter and J. T. Strong

*Central Electricity Generating Board,
Berkeley, UK*

INTRODUCTION

Since the initial development of the resistance strain gauge, users have attempted to make strain measurements at increasingly high temperatures. As indicated in other chapters of this book, a degree of success has been achieved over the years. For instance, successful measurements of dynamic strain can be made at temperatures up to the order of 1000 °C; static and quasi-static measurements can be made, with varying degrees of success, at temperatures up to the order of 550 °C.

The major limiting factors associated with high temperature resistance gauges are two fold:

(1)   Despite extensive investigations[1,2] it has not been possible to find a resistance alloy having basically stable and repeatable resistance–temperature characteristics to meet static and quasi-static requirements.

(2)   The materials suitable for bonding the resistance wires to the test structure, whether directly or indirectly, have limited insulation capabilities at the higher temperatures and this imposes a limit on dynamic as well as static capabilities.

Consideration of the principle of capacitance devices shows that such fundamental problems can be avoided. For instance, for a simple parallel plate capacitor

$$C \propto ak/t$$

where C is capacitance, a is area of plate, t is separation between plates and k is dielectric constant. Thus the capacitance can be varied by changing the area, a, or the gap, t. This means that capacitance devices can be made to depend only on geometrical features. Provided the capacitor terminals are isolated from earth, the electrical properties of the materials used are relatively unimportant, so they can be chosen to meet the mechanical requirements. This provides a significant advantage over resistance wires.

Until recently, the major problem delaying the development of capacitance gauges, has been that associated with the small size of device required by the structures analyst. Only small values of capacitance can be built into such devices and it has not been possible to measure these small values within acceptable levels of sensitivity and resolution. For example, in 1945 Carter et al.[3] reported their development of a capacitance strain gauge having a gauge length as small as 16 mm, but because of inferior measurement capabilities, conventional resistance gauge technology proved increasingly able to better the requirements and the capacitance gauges were abandoned.

The major problem of capacitance measurement was that the capacitance instabilities of the connecting cables, significant in terms of gauge capacitance, could not be isolated from the gauge signal. However, recent measurement technology using the 'three terminal method' has allowed these problems to be overcome. It is now relatively easy to measure gauge capacitance of less than 1pF with a sensitivity of the order of 10^{-4}pF and resolution of similar order.

Over the past 10–15 years there has been new impetus from the aerospace, power and process industries to measure static and quasi-static strains at higher temperatures and/or with better long-term stability than is possible with any available resistance gauges. This has led to the current situation where two types of capacitance gauges are available to the commercial user and a third type was available for a period of time. Each was developed to meet different specific requirements and is quite different in mechanical principle. The discontinued gauge used constant area parallel plates separated by mica. Both the currently available types rely on the air gap principle, but while one is strictly a variable gap parallel plate capacitor, the other uses variable area cylindrical plates.

These gauges are each in their own way unique and consequently each is discussed individually. So far as possible the background to each development is given and sufficient data presented to enable the reader to decide how the gauges can and have performed in their respective fields. The presentations are made with the commercial user in mind, so that he can judge the potential that the gauges might have to satisfy his own requirements.

It can be argued that the chapter should be restricted to discussing only the two types of gauge currently available since, because the third is no longer available, it can only be of little interest. The development work, however, is of considerable interest, not least because the gauge was required to measure in quasi-static situations to higher temperatures than any other known development.

HUGHES MODEL CSG-201 HIGH TEMPERATURE STRAIN SENSOR

First Developments

The first developments of the Hughes capacitance gauge began in 1966 at the Hughes Aircraft Company, to support the development and qualification testing of aircraft, missiles and space vehicles.[4] The required operating temperatures, at that time, achieved as a function of vehicle speed and altitude were predicted to be up to 800 °C.

The intended application of this gauge is a typical quasi-static requirement in that it was required to measure strains induced by aerodynamic forces and temperature changes over relatively short periods of time. This requirement was reflected in the target specifications by setting relatively close limits on gauge factor change with temperature and apparent strain due to temperature, but making generous allowance on high temperature drift rate.

A number of different gauge configurations were studied. One was a sawtooth design in which plate area and gap vary simultaneously. The others were either variable gap or variable area systems. The final choice was a variable gap system mounted in a rhombic frame. The rhombic frame, shown in Fig. 1, was chosen for several reasons:

(1) It acts as a strain intensifier. The ratio of gap change to applied strain is 2:1.

(2) The gap width decreases with increasing strain, thus capacitance increases with increasing strain.

(3) It was claimed to be a very flexible member requiring forces of the order of 1·3 to 2·3 kg to apply strains of 5000 μm/m.

The aim was to fit a capacitance element producing a capacitance value of between 10 and 100 pF. To achieve this, the elements were made from four stainless steel plates separated from each other and the rhombic frame by five mica dielectric insulators. The multi-plate design increased the

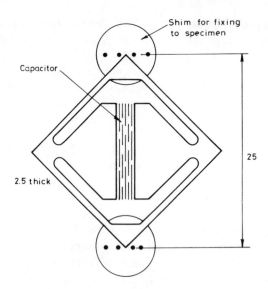

FIG. 1.   Rhombic frame.

effective plate area and the mica had a dielectric constant of about 4 at 21 °C (air = 1). Unfortunately, the dielectric constant of the mica increased with temperature in the manner indicated in Fig. 2, so for a constant gap, capacitance increased with temperature. However, it was considered that this effect could be offset, at least to a reasonable degree, by deliberately mismatching the thermal coefficients of expansion (TCE) of the rhombic frame and test specimen material. In this way, the plates would move apart as temperature increased, and counteract the increase in dielectric constant. A mismatch of about 2 ppm/°C was considered appropriate so, for the development stage, the rhombic frames were made from stainless steel 310 having a nominal TCE of 18 ppm/°C, for use on Inconel test bars, with a TCE of 16 ppm/°C.

Forty gauges were manufactured for evaluation. Considerable scatter occurred in most parameters but it was concluded that since each gauge could be calibrated prior to use, reliable data could be obtained from them. The main features are discussed in turn.

Gauge Capacitance
The room temperature, unstrained capacitance varied between 11·2 and 16·4 pF.

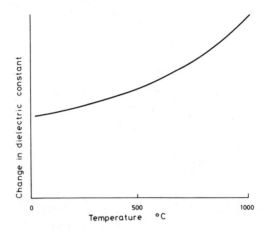

FIG. 2.   Effects of temperature on dielectric constant (typical for mica).

Gauge Factor
Collectively, gauge factor at room temperature varied from 19·5 to 40·5. Individually, gauge factor change with temperature was typically 30 % in the range 24 to 800 °C and the variations were not linear and varied differently for a gauge in tension and compression. Typical curves are shown in Figs 3 and 4.

Strain and Capacitance
At any one temperature capacitance change versus applied strain was relatively linear over the test range of 1000 μm/m. But the output at different temperatures reflected the change in gauge factor. This can be seen by comparing the output of the gauge plotted in Fig. 5 with the appropriate plot of gauge factor change in Fig. 3.

Drift Rate
Nine gauges were tested at 800 °C for a 1 h period at nominal zero strain. The drift varied between − 20 and + 125 μm/m over the full period. Any additional effects of strain on drift rate were not reported.

Zero Shift per Cycle
Fifteen gauges were temperature cycled. The maximum zero shifts occurred on the first cycle, the values ranging between − 100 and + 250 μm/m. On subsequent cycles the zero shifts were generally within ± 50 μm/m.

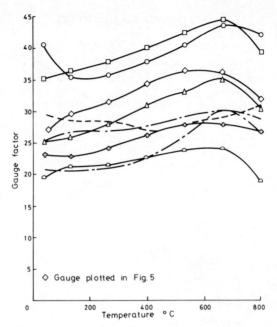

◇ Gauge plotted in Fig. 5

Fig. 3.   Gauge factor versus temperature (tension).

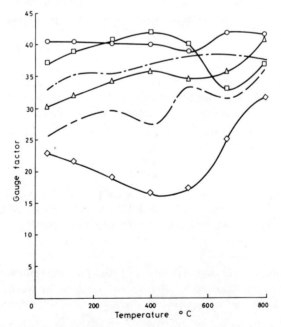

Fig. 4.   Gauge factor versus temperature (compression).

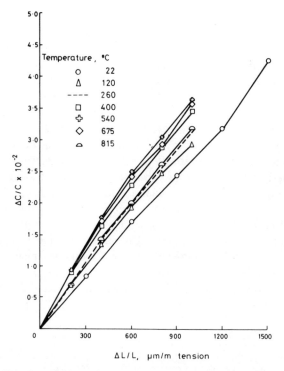

FIG. 5.   Change of capacitance versus applied strain.

Despite the variable performance on a gauge-to-gauge basis and the variations of gauge factor with temperature, it was considered that the development was relatively successful. Because the basic design allowed the gauge to be reused it could be calibrated in all aspects prior to use. By adopting this approach the developers claimed that quasi-static strains could be measured at 800 °C to within $\pm 10\%$. Clearly, the work involved in this approach was considerable, but since no alternative was available it was seen as a significant step forward in the field of high temperature strain measurement.

Second Stage Developments
Basic Design and Attachment
Presumably, as a result of the relative success of the earlier developments and the requirement to make measurements at still higher temperatures, Hughes Aircraft Company started the development of a strain gauge in

1970[5] which, subsequently, was made available to the commercial market as the model CSG-201 High Temperature Strain Sensor.[6] The development specification was aimed at a gauge suitable for making measurements at temperatures up to 1100 °C. The work was seen as a logical extension of that already described, to increase the temperature capability and to decrease the gauge length to 12·5 mm. A life of 30 mins at 1100 °C was specified.

The initial phases of the work reconsidered the choice of gauge configuration and the dielectric, with due regard to the increased temperature and smaller size requirements. The rhombic frame was retained but, in the new design, the gap for the capacitor was turned through 45°. This is shown in Fig. 6. The advantage claimed for this design was that it is more sensitive, i.e. the ratio of gap change to applied strain is 2·8:1 as opposed to 2:1 in the earlier design, and consequently more suitable for a smaller gauge. It also introduced a potential advantage in that the gauge could be mounted on the test structure with either diagonal in line with the applied strain. This means that if the gauge performed better in an increasing as opposed to a decreasing capacitance situation (or vice versa) then it could always be mounted to operate in its best direction. This supposes, of course,

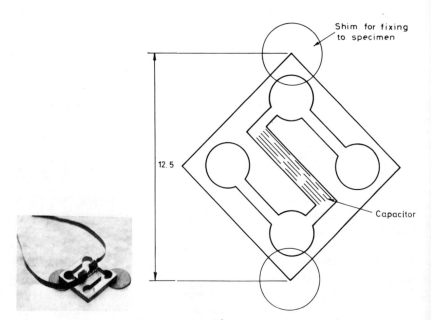

FIG. 6.   Rhombic frame No. 2.

that the applied strain direction could be predicted. Also, if used in this way, the advantage of knowing that increasing capacitance related to increasing strain would be lost.

The same basic four plate capacitor design, using mica as the dielectric, was also retained.

The method of attaching the gauges to the test structure again used shim discs spot-welded to the two corners of the rhombic frame. Although spot-welding these shims to the test structure was shown to be adequate and generally recommended, various bonding techniques were investigated. These were intended for use in situations where spot-welding might be considered detrimental to structural integrity or for bonding to ceramic coated surfaces. Flame-sprayed ceramic (Metco 450) proved highly successful on metal surfaces and a ceramic adhesive was recommended for bonding to ceramic coatings.

Performance

As with the gauges developed earlier, there was considerable scatter in initial capacitance and room temperature gauge factor. This is, perhaps, understandable from consideration of the dimensions and construction of the gauge. For instance, the four plates were nominally 5 mm long, 2·5 mm high, separated and insulated from the frame with mica within a 1 mm wide strain frame gap. Normal tolerances on these dimensions, particularly including the thicknesses of the mica layers and the consequent compression of these when inserted in the gap, will have significant effects on both capacitance and gauge factor. Also, the gauges had to be given a thermal cure on completion of manufacture and test results show that this had significant effects on the subsequent performance. It is not surprising, therefore, that in a batch of 25 gauges, scatter on initial capacitance was $\pm 34\%$ of the average value and room temperature gauge factors varied between 22·2 and 69·5. Also, gauge factors of individual gauges varied between tension and compression by over 30% in some cases.

From the foregoing, it is clear that one of the claimed major advantages of the gauge, that it could be repeatedly reused, was in fact, essential. Each gauge had to be calibrated before use in order to determine its basic features and the effects of temperature and strain on these. Further, because of this approach, it was essential to know the repeatability of performance when remounting the gauge on the test structure after removal from its calibration site. For this reason, in the subsequent presentation of a sample of test results, repeatability is emphasised.

Figures 7 and 8 show the run-to-run variation in gauge factors. The data are presented as envelopes encompassing run-to-run deviations of all gauges tested within the given temperature range. In Fig. 7 the heavily shaded area included the results of 10 gauges cured at 880 °C. Above 950 °C the gauge outputs became so erratic as to be immeasurable. The lightly shaded area includes results from 5 gauges cured at 1100 °C. In Fig. 8 the results were obtained from both groups of gauges. From these results, there are two points worth noting:

(1)   The gauges most definitely perform better in compression rather than in tension. This suggests that advantage could be taken of the ability to mount the gauge on the diagonal which would induce the capacitor to operate in the best direction independent of the applied strain. This presumes, of course, that the applied strain direction is known and that the gauge will have sufficient strain range in its nominal 'compressive' directions.

(2)   Although the higher curing temperature resulted in data being obtained at higher temperatures, the performance of these gauges throughout the full temperature range was worse than the performance of those cured at the lower temperature.

Figure 9 shows actual change in gauge factor against temperature for a single gauge in tension and compression. Similar plots for a number of gauges in tension only are shown in Fig. 10.

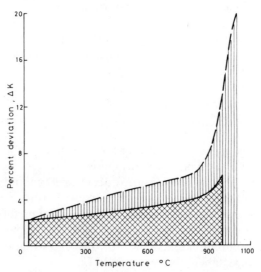

FIG. 7.   Repeatability, run-to-run (tension).

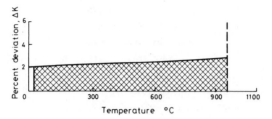

FIG. 8. Repeatability, run-to-run (compression).

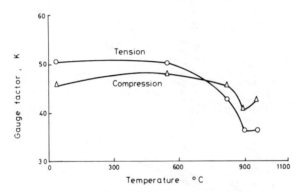

FIG. 9. Gauge factor versus temperature (tension and compression).

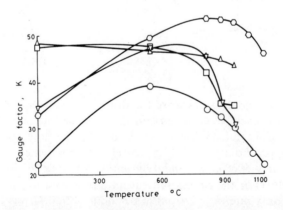

FIG. 10. Gauge factor versus temperature (tension).

Figure 11 shows a typical gauge output at 1100°C for five successive strain cycles between zero and 0·1% strain. The difference in outputs between increasing and decreasing strain was predicted to be due to drift. The parallel broken lines drawn through the maximum strain point of each cycle indicate the degree of cycle-to-cycle repeatability ignoring the drift. Since these lines pass approximately through the mid-point of the two zero readings on each cycle (ignoring the first cycle) repeatability is fairly good.

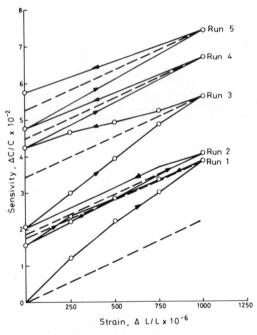

FIG. 11.   Gauge sensitivity.

Figures 12 and 13 summarise the drift test results. These are, again, presented as envelopes encompassing the range of data obtained from the group of gauges tested in each case. Assuming an average gauge factor of 35, the upper boundaries of the envelopes represented drift rates of approximately 6000 μm/m and 1200 μm/m per hour in Figs 12 and 13 respectively. (For the gauges subsequently produced for the commercial market a drift rate of <250 μm/m per hour at 950°C was claimed. No evidence has been published as to how this improvement was achieved.)

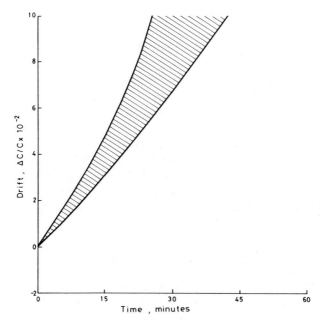

FIG. 12.   Drift at 1100 °C (gauges heat cured at 1100 °C).

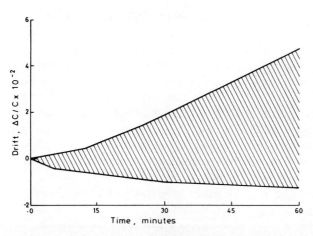

FIG. 13.   Drift at 950 °C (gauges heat cured at 885 °C).

Much more data than presented here has been published by the gauge developers. Nevertheless, sufficient data is presented to indicate the extensive scatter in capacitance and gauge factor and the effects of temperature and, in some cases, strain on these parameters. It was concluded by the developers that the gauges were capable of yielding reliable measurement data up to about 950 °C. These conclusions were based on the degree of repeatability of individual gauges, claimed to be less than 5 %. To retain this repeatability the working ranges were restricted to $+0 \cdot 5$–$0 \cdot 2 \%$ strain at room temperatures reducing to $\pm 0 \cdot 2 \%$ at 950 °C.

The most limiting feature of the gauges proved to be the high drift rates. At best, therefore, the gauges could only be used reliably for very short-term measurements at constant temperature or in relatively short-term thermal transient situations and to meet these requirements considerable precalibration was essential.

It is, of course, easy to criticise the apparent poor performance of these gauges, as compared with lower temperature devices. But it should be remembered that no other gauge developments for such high temperature (quasi) static applications have been openly published. The gauges were developed to meet a specific requirement and only their suitability to meet that requirement was investigated. There is, therefore, no direct comparison of performance with lower temperature devices. It can only be presumed that their inability to find a commercial market was due to inadequate performance standards.

THE BOEING/HITEC STRAIN GAUGE

This gauge resulted from a decade of development jointly funded by The Boeing Company and the NASA Flight Research Centre in the USA.[7] Development work started at about the same time as Hughes were beginning their work on the gauge just described, but the basic requirements were more broadly based. The same operating temperature of 800 °C was set, but in addition to the envisaged transient and mainly short-term applications within their own industry the developers also considered longer term high stability requirements of other industries and the nuclear power industry in particular.

Whereas the Hughes gauge used mica as the dielectric material to increase gauge capacitance, Boeing took the view that the dielectric should be as stable as possible and hence an air gap system was chosen. Over the temperature range to 800 °C the change in dielectric constant of air is about 4%, but because the resulting capacitance and, in particular, its change with strain is very small, special measuring circuitry was developed in parallel with the gauge. (For the final developed system the total capacitance is nominally 1 pF and capacitance change with strain is approximately 0·018 pF/1000 μm/m.)

The Gauge

A section of a gauge is shown in Fig. 14 and a schematic view in Fig. 15. The major components are:

(1) A compensating rod, usually made of the same material as the test specimen. The rod establishes the gauge length and provides nominal self-temperature compensation.
(2) A pair of cylindrical excitation plates, mounted on, but electrically insulated from, the compensating rod.
(3) A sensing ring, coaxial with and surrounding the excitation plates.
(4) Attachment ribbons for fixing the gauge to the test specimen.
(5) Flexures to maintain coaxial alignment of the excitation plates and sensing ring.

As the specimen dimensions change due to strain the outer cylinder moves axially with respect to the two inner cylinders, changing the differential capacitance. In principle the gauge length can be made to any desired dimension by simply changing the length of the compensating rod.

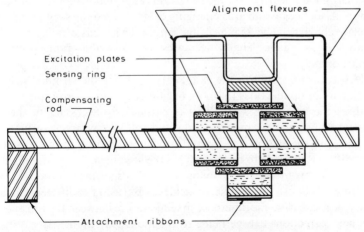

FIG. 14.   Cross section of Boeing/Hitec gauge.

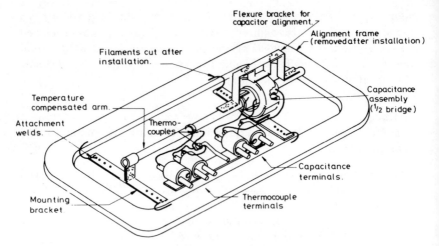

FIG. 15.   Schematic of Boeing/Hitec gauge.

They were developed originally with a gauge length of 25 mm but are now also supplied with a gauge length of 6 mm.

The spot-welds used to fix the attachment ribbons to the specimen are not made within about 4 mm of the ribbon centre. This, together with the flexure bracket, maintains concentricity of the excitation plates and sensing ring. Thus changes in capacitance result only because more or less area of the sensing ring overlaps the respective excitation rings. This leads to extremely good linearity over the full strain range in both compression and tension. Also, because an area changing device is used the strain range can be relatively large. The 25 mm gauge length device has a range of $\pm 2\%$ strain and as gauge length decreases the available range increases proportionately.

Electrical connections to the gauges are made via the terminal strip. In addition to the connections for the capacitance elements, provision is made for three thermocouple wires. This is particularly necessary in thermal transient situations. Because the rod stands off the specimen surface, the differences in thermal mass, draughts and uneven heating mean that its temperature is unlikely to be the same as the specimen. Thus, although the rod is made from the same material as the specimen, exact thermal matching is almost impossible to achieve. By using a differential thermocouple to measure the difference in temperature between the rod and the specimen corrections can be made to the indicated strain.

The specially developed instrumentation consists of a mode card which is used in conjunction with a commercial signal conditioner/power supply. The mode card applies an a.c. excitation signal to the gauge and converts the gauge output to an analogue voltage. This output voltage ranges between ± 5V d.c. and is suitable to drive any voltage measuring device such as a chart recorder, plotter, oscilloscope or DVM. Calibrated controls for capacitance balance and gain permit the d.c. outputs to be converted directly to strain. A signal conditioner and mode card module is required for each gauge being used.

Installation

The gauge must be mounted on a flat surface, but tapered shims can be used under the gauge feet to provide a flat mounting when the specimen surface is curved.

Each gauge is supplied in a mounting frame, as shown in Fig. 15, which holds the gauge at its correct gauge length. The mounting ribbons are welded to the prepared specimen surface using a single row of spot-welds, but not closer than 4 mm to the gauge centre line. The plate holding the terminal strip is attached to the surface by a single spot-weld and then the mounting frame is removed.

Performance

To date, the Boeing/Hitec gauge has not been used in the UK and little performance data has been published in open literature by users. Therefore, the data presented here is obtained, in the main, from Harting[7] and the gauge manufacturers.[8] Where possible, this is supported by data from other sources.

GAUGE CAPACITANCE, LINEARITY AND GAUGE FACTOR

The overall capacitance of a correctly mounted gauge is approximately 1 pF and gauge sensitivity, which is the important measurement parameter, is nominally 0.018 pF$/1000\,\mu$m/m strain. Due to normal manufacturing tolerances each gauge will vary about these nominal values. Thus each gauge requires precalibration which is done by the manufacturer.

Because linearity of output with applied strain is very good the calibration data is used to set the gain of the signal conditioning equipment so that output correlates directly with strain. Typical graphs of gauge output against applied strain at various temperatures are shown in Fig. 16.

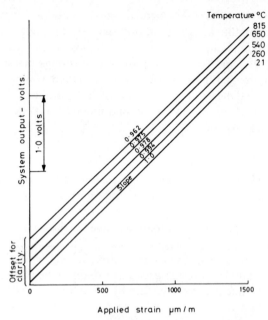

FIG. 16.   Gauge output due to strain.

These are, clearly, linear and the change in slope between room temperature and 800 °C is less than 5 %. In other words, gauge factor does not change with strain but reduces with temperature by <5 % up to 800 °C which is in line with the change of dielectric constant.

Calibration tests by the developers to higher strain levels indicate that, for a 25 mm gauge, linearity remains within 0·5 % to ±0·3 % strain and within 2 % to ±2 % strain.

The effectiveness and importance of the flexure plates in maintaining concentricity between the excitation plates and sensing ring is shown in Figs 17 and 18. Figure 17 shows the output of a gauge in pure bending, on a 4·75 mm thick beam. The strain is overestimated by 11 %. Assuming the error is linear with specimen thickness it can be taken as 5·2 % per 10 mm of specimen thickness. In many applications the bending component will be unknown, thus it will not be possible to make corrections. Figure 18 shows the importance of taking great care not to damage or distort the flexure plates. Sensitivity changes significantly (100 % in the extreme) if good concentricity is not maintained. However, checks on condition can be made via the instrumentation. In a truly concentric device, the overall capacitance will remain nominally constant independent of strain. Therefore,

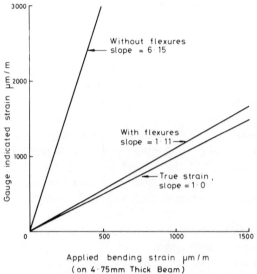

FIG. 17. Response of gauge to bending strain.

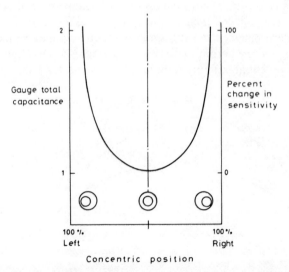

FIG. 18. Gauge sensitivity versus plate concentricity.

change in overall capacitance, which can be measured, indicates eccentricity.

Thermal Output

In general, thermal output of any strain gauge will depend on the relative properties of the gauge and specimen material. If the compensating rod and the gauge are made from identical material, having the same thermal coefficient of expansion (TCE), and the temperature of the rod and the gauge are always identical then the thermal output should be zero. In practice it is virtually impossible to achieve this state because both requirements are difficult to achieve. The differential thermocouple is provided for making corrections for temperature differences, but little can be done to guarantee an exact match of TCE.

Figure 19 shows gauge output with temperature change for a nominally matched gauge and specimen. At the lower temperature the apparent strain output is approximately $1\,\mu m/m/°C$ and also there is about $100\,\mu m/m$ difference between increasing and decreasing temperature. Therefore, not only is great care required with the choice of compensating rod material, but the hysteresis implies that some thermal output will arise in the best matched gauges.

Long-Term Drift

Figure 20 shows a number of curves of drift against time at 565 °C, from various sources.[7,9,10] The curves from references 9 and 10 were obtained from standard production gauges and consequently may be more typical than those obtained from development gauges.[7] Clearly the gauge has significant potential for long term static measurements.

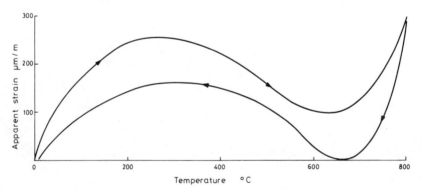

FIG. 19.   Boeing/Hitec gauge thermal output.

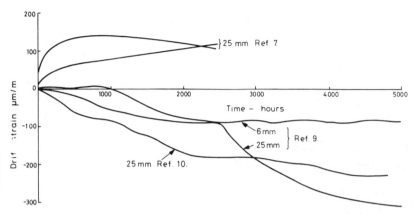

FIG. 20.   Boeing/Hitec drift versus time at 600 °C (on stainless steel specimens).

Thermal Cycling

It has not been possible to find published data indicating zero shift due to repeated temperature cycling but some data is available indicating repeatability of output due to load cycling. Elliot[10] shows the output of one gauge at 593 °C load cycled eight times over a strain range of 235 μm/m, with 100 h hold times between each load cycle, to repeat to within $\pm 2\%$. Smith[9] shows results from one 25 mm gauge and one 6 mm gauge held at 593 °C for 11 000 h and given two temperature cycles and five load cycles within this period. For both gauges, the strains indicated as a result of the applied load are within 2 % at the respective temperatures. For the 25 mm gauge the average of the three load applications at 593 °C is 2 % lower than the average at 24 °C. For the 6 mm gauge, the average is less than 5 % lower at 593 °C than at 24 °C.

THE CERL/PLANER CAPACITANCE GAUGE

Advances in the design of turbo-generator machinery for use in the power generation industry created the need for two specific types of strain measurement at temperatures up to 600 °C. One requirement was to measure thermally induced strains which occur during load transients, particularly during the daily start-up of the plant when the generator is loaded from zero to maximum within 30 to 60 mins. The second was to measure long-term creep strains at the maximum plant operating temperatures.

In the thermal transient situation resistance gauges proved reasonably satisfactory up to approximately 550 °C, but did not have the high stability capability for measuring long-term creep strains. In consequence, Noltingk et al.[11] considered the use of capacitance methods and developed a gauge, with long-term stability as one of the major requirements.

In order to appreciate the degree of stability required when attempting creep strain measurements the following parameters apply. The basic creep process of high temperature structural steels has three phases as shown in Fig. 21. In a well designed and correctly operated structure, creep strains in membrane regions may be of the order of 0·3 % in 10^5 h. This is an average

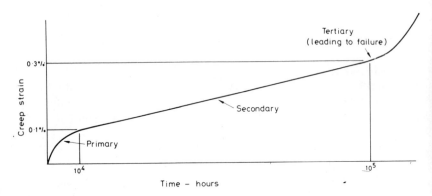

FIG. 21.   Idealised creep behaviour.

strain rate, from time zero, of 0·03 μm/m/h. The primary creep strain is generally considered to be about 30 % of the 10^5 h strain and to occur over about 10 % of the time, which is 0·1 % strain in 10^4 h or an average strain rate of 0·1 μm/m/h.

In discontinuity regions of particular interest stresses may be two to four times the membrane levels and the creep strain rates will range up to an order of magnitude greater than in the membrane areas. Further, in laboratory experiments steps can be taken to accelerate creep rates by up to an order of magnitude. Thus for useful creep strain measurements to be made, the strain gauges are required to be stable to at least the same order as the predicted membrane creep rates on real structures. That is, to make measurements during the primary creep phase, a gauge stability better than 0·1 μm/m/h is required and this should be improved to better than 0·03 μm/m/h if secondary phase measurements are to be worthwhile.

Gauge Development

For the Boeing/Hitec gauge an air gap device was chosen primarily because the dielectric constant of air is only marginally affected by temperature changes and consequently the same linear gauge output due to strain changes would be retained throughout the working temperature range. For the CERL/Planer gauge an air gap device was chosen primarily to provide stability. The only other similarity between the Boeing and CERL developments was that of developing special circuitry to measure the small levels of capacitance inherent in such air gap systems. Whereas Boeing chose to use a half bridge device having concentrically mounted circular plates, Noltingk *et al.* chose to develop a quarter bridge device having a simple rectangular plate variable gap capacitor. Then, in order to minimise the electrical precision required to measure capacitance changes at these low levels they sought to provide mechanical amplification between the strain applied and gap distance. In this respect, the approach was similar to that used in the development of the Hughes gauge, but the solution was totally different.

A number of configurations were considered leading to the ultimate choice shown in Fig. 22. This consists of two arched strips mounted one above the other. One capacitor plate is fixed under the top arch and the

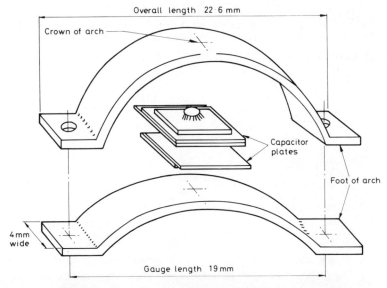

FIG. 22. Exploded view of gauge.

other is fixed on top of the lower arch, leaving a small air gap between the two. The arched strips are 4 mm wide and 0·1 mm thick thus only very small axial forces are required to deflect them and hence only a single spot-weld is required at each end for attachment to the test specimen. When axial strain is applied to the gauge the lower arch deflects by a greater amount than the upper and consequently the air gap and capacitance changes.

The changes can be calculated for any given instantaneous height of arch. A theoretical curve resulting from such calculations is shown in Fig. 23. It is seen from this, that the capacitance–strain change relationship is not linear and capacitance decreases with tensile strain. With modern data handling methods for converting from capacitance to strain, these are not considered to be disadvantages. The maximum available strain range is determined in the compressive mode by the capacitor plates touching and in the tensile mode by the lower arch tending to a flat strip. The gauge is manufactured such that in its free unmounted form it has approximately equal range in the two directions of at least $\pm 0.5\%$ strain. On attachment to a structure the gauge can be offset from this 'free' position to give the full 1% measuring capability in the chosen direction.

Since the capacitor works on the air gap principle there will be small changes in dielectric constant which affect its output, due to changes in pressure and temperature of the surrounding environment. However, at any given set of conditions the dielectric constant will not vary. The stability of the gauge, therefore, depends on the materials used in its construction and the way these behave in use. Creep stability of the component parts of the gauge is the main consideration.

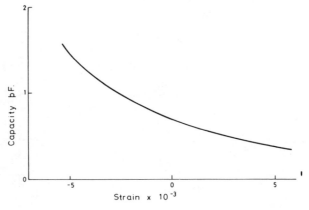

FIG. 23.   Predicted capacitance versus strain.

The maximum stressed parts are the attachment welds and the heels of the lower arch. One of the advantages of the arched strips is that only very small forces are required for their operation. The end force required to produce 0·5 % tensile strain, which is the worst case, is < 300 grams thus the stresses and hence strain, in the attachment welds should always be significantly less than that being measured. Also the bending strains at the heels of the lower arch are only about 50 % of that being measured. Taking this into consideration, by choosing suitable alloys for the arches it should be possible to make the gauge framework virtually totally creep resistant up to temperatures at least of the order of 750 °C. Nimonic alloys are used for this purpose. One type of gauge has a thermal coefficient of expansion (TCE) of approximately 11 ppm/ °C, for use on ferritic steels and the other approximately 17 ppm/ °C for use on austenitic materials.

Platforms for the capacitor plates are welded to the arched strips and these welds are, clearly, unstressed. The platforms are coated with alumina insulation and the platinum capacitor plates are bonded to these by a special technique. Lightweight coiled ribbon leads are attached to each capacitor plate for connection to the connecting cables.

Measuring Circuit
The basic measuring circuit, suitable for any quarter bridge capacitance gauge is shown in Fig. 24. As stated in the introduction to this chapter, the problems of measuring small capacitance levels and changes are created by the capacitance to earth of the cables required to connect the gauge to the measuring instrument. The cable capacitance will be relatively large, and vary from gauge to gauge and, more importantly, the capacitance of any one cable will vary with time and temperature, so that balancing techniques cannot be used to eliminate it. In the circuit shown, the shunt cable impedances appear as loads across both the oscillator and detector transformers, thus the cables affect both standard and unknown capacitors

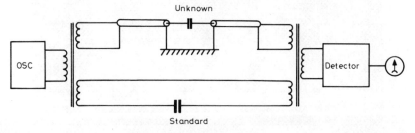

FIG. 24. Measuring circuit diagram.

equally. In principle, therefore, the value of gauge capacitance is obtained by comparison with a standard. Clearly, the quality of the instrumentation is important, to attain the required sensitivity, stability and linear response. To this end, a suitable bridge and multi-plexing system were developed in parallel with the gauge and are available commercially in various forms to suit users' requirements. With this equipment it is possible to measure capacitance with a repeatability better than 10^{-3}pF which, with this gauge, is better than 1 μm/m.

Installing the Gauge

The gauge is normally installed on a clean flat surface, which has been prepared in a similar manner to that required for conventional electric resistance gauges. An installed gauge is shown in Fig. 25. The two stainless steel sheathed, mineral insulated cables with suitable high temperature end seals, are fixed in position. One end of the gauge is then spot-welded to the surface and the flying leads from the gauge electrodes spot-welded to the cores of the coaxial cables. By connecting a measuring instrument into the circuit it is now possible to monitor the gauge output. The free end of the gauge can be moved until a reading appropriate to the desired gauge preset is indicated, when the final spot-weld can be made.

Any standard capacitance discharge spot-welder is suitable for making

FIG. 25.   Gauge installation on pressure vessel.

the attachment welds. The required weld power of about 12J is determined by tests on samples of the gauge material.

Performance

Development work to improve gauge performance has been almost continuous since the first production gauges were made. Although the early production gauges were highly stable in their zero strain condition they did not exhibit the same degree of stability as strain was applied. This led to changes in material and heat treatment and improvements in construction methods to meet production requirements, culminating in gauge types C2 for use on ferritic steels, and C3, for use on austenitic steels. The C2 gauges in particular were manufactured in considerable quantities and used fairly extensively within the power generation industry. Numerous performance evaluations were carried out and examples are given in subsequent sections of this chapter.

Although the long-term stability of the C2 and C3 series gauges met the standards required as outlined on page 312, they did not perform consistently in thermal cycling situations. The problem was traced to a thermal ratchetting phenomenon of the material used to bond the capacitance plates to the mounting platforms, leading to 'zero shift' on each completed temperature cycle. Modifications made to overcome the problem resulted in the currently produced gauges, designated C4 and C5 for use on ferritic and austenitic materials, respectively. These gauges perform identically to the earlier C2 and C3 series in all aspects except thermal cycling, which is considerably improved. Thus all the results presented are applicable to the currently available C4 and C5 series gauges.

Gauge Factor

The relationship between capacitance and applied strain is shown in Fig. 23. Dimensional tolerances in production lead to an overall scatter band of about $\pm 15\%$, thus each gauge requires calibration before use. The calibrations are normally provided by the supplier, but the user can obtain the required equipment to make his own calibrations if he desires. The individual curves are adequately described by a sixth order polynomial for computerised data handling.

Due to the change in dielectric constant with temperature, of the order of 4 to 5 % over the temperature range to 750 °C, there is a similar change in gauge factor. This is not of concern when measurements are made at constant temperature since capacitance change due to strain remains proportionally the same. When measuring in thermal transient situations the user can make corrections if he thinks it worthwhile.

Long-Term Stability

As with the Boeing/Hitec gauges all known users for creep strain applications are concerned with temperatures up to 600 °C. Therefore all test data has been obtained at this nominal temperature.

Manufacturers' literature quotes drift rates of C4 gauges on Nimonic 80A test specimens of <50 μm/m in the first 100 h reducing to 0·03 μm/m/h in the long-term and this is confirmed by other workers. Because of the requirement to make measurements on ferritic steel pressure vessels, $2\frac{1}{4}\%$ Cr 1 % Mo in particular, tests were made on this material by the authors. Figure 26 shows typical results from a test on Nimonic 80A and a test on

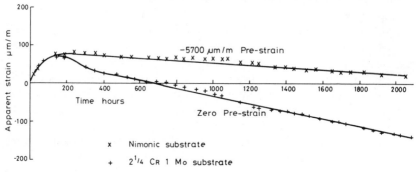

FIG. 26.   Long-term stability of C2 gauges in air at 600 °C.

$2\frac{1}{4}\%$ Cr 1 % Mo. The behaviour of the gauge on the Nimonic specimen supports the manufacturers claims. That on the ferritic specimen is significantly worse; positive drifts of up to 100 μm/m were recorded in the first 200 h followed by negative drifts of approximately 0·1 μm/m/h. It was noted that a considerable build-up of oxidation had occurred on the ferritic specimen (Fig. 27) whereas on the Nimonic specimens this did not, of course, occur. Therefore, it was suspected that a slow build-up of scale under the welded feet of the gauges had progressively distorted the arches resulting in the higher drift rates.

Using C2 gauges, tests were made on ferritic steel specimens in an inert atmosphere to prevent the oxidation. The results shown in Fig. 28 are typical and in line with those obtained on Nimonic specimens and confirm that oxidation of ferritic specimens needs to be prevented if best

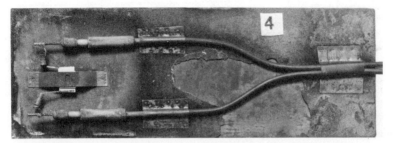

FIG. 27. Oxidation growth on $2\frac{1}{4}\%$ Cr 1% Mo specimen.

performance is required from the gauges. The results obtained were summarised by Procter and Strong[12] as follows:

(1) Total drift during the first 200 h, 50–70 μm/m.
(2) Drift per hour from 200 to 1200 h, 0·1 μm/m.
(3) Drift per hour from 1200 to 3000 h, 0·02 μm/m reducing to 0·005 μm/m.

The subsequent modifications to the gauge to improve the performance during temperature cycling, culminating in the C4 and C5 gauges, involved a change in the plate to platform bonding technique. Unfortunately, although this virtually eliminated the zero shift problem the bond is not effective in reducing atmospheres and therefore the gauges cannot be used in inert gases. Probably the easiest solution to overcome the oxidation

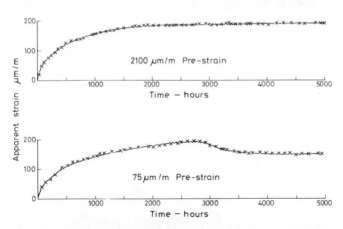

FIG. 28. Long-term stability of C2 gauges in argon at 600 °C.

FIG. 29.   Gauge mounted on shim carrier.

problems is to spot-weld a flat stainless shim to the ferritic surface, then
mount the gauge on the shim. A photograph of such an installation is
shown in Fig. 29, and the corresponding long-term drift behaviour
in Fig. 30, which is clearly of the highest standard. An attractive
solution may be to locally plate the specimen surface at each strain gauge
position and spot-weld the gauge on to this. It would, of course, be
necessary to choose a suitable plating composition. The authors made tests
using gold plating, since this was readily available to them. Although the
plated surface remained sound and did not oxidise at the test temperature,
the gauge performance was clearly inferior to those mounted on untreated
surfaces. This was most probably due to the attachment welds having
inferior creep resistance when made on the gold plate. It is possible that
nickel, or similar plating, will satisfy the requirement, but this has not, so
far, been investigated.

Thermal Cyclic Behaviour
When using the gauges in thermal cycling situations two aspects are of
concern to the user; (i) the apparent strain output on each cycle and (ii) the

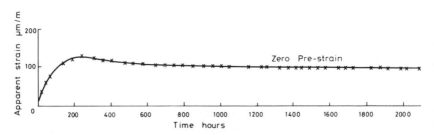

FIG. 30.   Long-term stability of gauge on shim carrier at 600 °C.

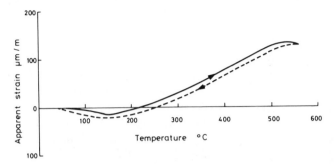

FIG. 31. Thermal output of C4 gauge.

repeatability of output during successive cycles. Both aspects are shown in Figs 31 and 32.

Figure 31 shows a single cycle output of a type C4 gauge on a ferritic specimen during heating and cooling at approximately 50 °C per hour.[13] The general shape of the curve will clearly relate to the relative coefficients of expansion of the gauge and specimen material. The important aspect to note is that the heating and cooling curves are almost identical and the zero shift on completion of the cycle is less than 10 μm/m.

Figure 32 shows repeated outputs at 600 °C and room temperature[14] from one C4 and one C5 gauge, for 50 consecutive cycles as follows: heat to 600 °C in 4 h, hold for 4 h and then cool to room temperature in 16 h. Thus, these curves indicate the degree of zero shift per cycle plus any drift occurring during the times at high temperature. In the worst of the two cases the shift per cycle is less than 10 μm/m per cycle.

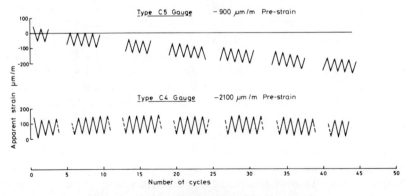

FIG. 32. Cyclic drift of CERL/Planer gauge.

OVERALL CONSIDERATIONS

As already stated, the Boeing/Hitec and the CERL/Planer gauges are the only two general purpose capacitance strain gauges available on the commercial market. Except that both gauges are designed around an air gap system, both are quite different in working principle. But it is the performance capabilities which are of prime interest to the potential user and these have been discussed in varying detail in the preceding pages. However, for easy reference, appropriate extracts from the two manufacturers specifications, are shown side by side in Table 1.

Clearly, each gauge has advantages over the other. The Boeing/Hitec gauge can be made with a gauge length as small as 6 mm and it has between two and eight times the strain range (depending on gauge length) of the CERL/Planer gauge. On the other hand the CERL/Planer gauge appears to be potentially more stable in the long-term and is significantly less sensitive to bending. The specifications also show the gauge sensitivity (capacitance change per unit strain) of the CERL/Planer gauge to be considerably higher than that of the Boeing/Hitec gauge. Because both gauges have been designed to be compatible with their own individual capacitance measuring systems this is not necessarily significant.

Further, gauge performance specifications are not the only consideration when selecting a gauge. In many situations both gauges will undoubtedly meet users requirements and the choice of gauge to be used will be influenced by cost, availability, users or reported experience from previous applications and even salesmanship. To date, there is insufficient information to allow comparisons of these practical aspects thus only a few general points can be made.

The Boeing/Hitec gauge must be mounted on a flat surface thus wedge pieces must always be made in any other situation, whereas a number of users have mounted CERL/Planer gauges, with apparent success, on pipes and cylinders down to 150 mm radius. Both gauges are, clearly, fragile and appropriate care is required when handling and fixing. Both gauges will be similarly affected by environmental influences such as dust and humidity contamination of the plates. Protective covers can be used over the gauges which are probably easiest to apply in the CERL/Planer case, but will not provide 100% protection in either case.

Quartz insulated lead wires are generally recommended for the Boeing/Hitec gauges whereas mineral insulated sheathed cables are always used with the CERL/Planer gauges. The former are probably more susceptible to damage than the latter and are certainly more susceptible to humidity at

TABLE 1
COMPARISON OF SPECIFICATIONS

	Boeing/Hitec		CERL/Planer	
	DC-100	DC-025	Type C4	Type C5
Gauge length	25 mm (1 in.)	6 mm (0·25 in.)	19 mm	19 mm
Expansion coefficient	To match customers material		11 ppm/°C	17 ppm/°C
Strain range	±2%	±8%	1%	1%
Temperature range	Cryogenic to 800°C		Cryogenic to 750°C	
Output	Linear		Non-linear	
Typical sensitivity (capacitance change for 0·1% strain)	0·018 pF	0·0045 pF	0·1 pF (av)	
Drift at 600°C:	Calibration supplied		Calibration supplied	
	(material not quoted)		(on Nimonic 80A)	(on SS)
Initial			<50 με in 100 h	<10 με in 100 h
Short-term	<0·1 με/day		<0·1 με/h (av)	<0·1 με/h (av)
Long-term			<0·03 με/h (av)	<0·02 με/h (av)
Temperature cycle drift 600°→20°→600°C	None given		Typically 5–10 με/cycle	
Bending sensitivity on 10 mm thick specimen	5·2% (obtained from Fig. 17)		2%	

low temperatures. In both cases, care is needed when making the spot-welded connections to the lead wires to avoid shorts or open circuits.

Also, in both cases, capacitance coupling has been observed when unshielded lengths of excitation and signal leads, on adjacent gauges, are in close proximity to each other. Unshielded leads should, of course, be as short as possible. But if the problem can still not be avoided, foil guards can be used as shields.

Presumably because of the relative infancy of these capacitance gauges there is a marked lack of information in the open literature reporting on the success, or otherwise of applications of the gauges. Procter and Strong[12] reported the aforementioned evaluation tests in preparation for a pressure vessel creep test and Procter[15] subsequently reported on the performance of the gauges during the test. This latter paper is of interest in that it indicates the type of measurement problems which can arise during structural testing despite all the previous evaluation and proving exercises and which are not a direct result of gauge behaviour. Roland[16] reports applications of the Boeing/Hitec gauges at 590 °C to investigate thermal ratchetting of pipes and elastic/plastic/creep behaviour of shear lag specimens.

The authors have successfully completed a second pressure vessel creep test using CERL/Planer gauges and are aware of numerous site investigations on power plant components in which the gauges have been used with success to establish fault conditions. Two similar applications of the Boeing/Hitec gauge are briefly reported by Roland[16] and presumably there are other unreported instances of success with these gauges.

ACKNOWLEDGEMENT

This paper is published by permission of the Central Electricity Generating Board.

REFERENCES

1. BERTODO, R., Resistance strain gauges for the measurement of steady strains at temperatures above 650 °C, *J. Strain Anal.*, 1965, **1** (No. 1), 11–99.
2. BEAN, W. T., Res. program to develop alloys for up to 1200 °F strain gauges, *NASI-4644, NASA-CR-66775, N69-22187*, March 1969.
3. CARTER, B. C., SHANNON, J. F. and FORSHAW, J. R., Measurement of displacement and strain by capacity methods, *Proc. Inst. Mech. Eng.*, 1945, 215–21.

4. GILLETTE, O. L. and VAUGHN, L. E., Research and development of a high temperature capacitance strain gauge, *US Air Force Flight Dynamics Lab. Tech. Report, AFFDL-T12-68-27*, Feb. 1978.
5. GILLETTE, O. L., Research and development of a high temperature capacitance strain gauge for operations at 2000°F, *US Air Force Flight Dynamics Laboratory Report AFFDL-TR-11-103*, Oct. 1971.
6. Hughes Aircraft Company, Product data sheet, Model CSG-201 High Temperature Strain Sensor, 1972.
7. HARTING, D. R., Evaluation of a capacitive strain measuring system for use up to 1500°F, *Proc. ISA. Int. Symposium*, Philadelphia, USA, May 1975.
8. Hitec Corporation, Nardone Industrial Park, Westford, USA.
9. SMITH, J. E., Oak Ridge National Laboratories, Tenn., USA, Private communication.
10. ELLIOT, D. N., Westinghouse Electric Corporation, Madison, USA, Private communication.
11. NOLTINGK, B. E., McLACHLAN, D. F. A., OWEN, C. K. V. and O'NEILL, P. C., High stability capacitance strain gauge for use at extreme temperatures, *Proc. IEE*, **119** (No. 7), July 1972.
12. PROCTER, E. and STRONG, J. T., High temperature creep strain measurements using a capacitance type strain gauge. *Proc. Transducer'77 Conf.*, Wembley, England, July 1977.
13. WILLIAMS, J. A., CEGB, Marchwood Engineering Laboratories, Private communication.
14. PHILLIPS, L. S., G. V. Planer Ltd, Private communication.
15. PROCTER, E., High temperature creep measurements using a capacitance type strain gauge. *Proc. BSSM Conf.*, University of Bradford, England, September 1978.
16. ROLAND, D. M., An overview of capacitance strain gauges in the United States, *Proc. 6th Int. Conf. on Experimental Stress Analysis*, VDI-Berichte Nr. 313, Munich, West Germany, September 1978.

Chapter 8

THE VIBRATING WIRE STRAIN GAUGE

I. W. HORNBY

Central Electricity Research Laboratories, Leatherhead, UK

INTRODUCTION

The vibrating wire (VW) strain gauge consists of a thin steel wire held in tension between two anchorages (Fig. 1). The wire is set into transverse vibration by exciting it with a short pulse of current passed through the coil of an electromagnet positioned near the mid point of the wire. The same

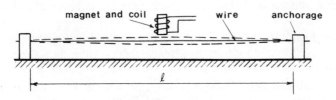

FIG. 1.   Vibrating string and electromagnet ($Q \propto l^2$).

coil is then used to detect the frequency of the vibrating wire. When the distance between the anchorages changes, the tension of the wire and its natural frequency also change. This form of gauge has been shown to be stable over periods exceeding 15 years and it has the advantage that its frequency is not changed by the resistance or length of the leads. Gauges of this design are used for measuring strain from a variety of materials and structures but it has been found ideally suited as a gauge which can be cast or embedded into concrete.

HISTORY

The vibrating wire principle of strain measurement has been known for many years and reports of its use have appeared as far back as 1928 in a paper by Davidenkoff.[1] This was followed by papers in France, Germany and the UK. The initial experiences in the UK were gained at the Building Research Station (BRS)[2,3] and then later by the Road Research Laboratories.[4]

These early gauges were predominantly for measuring surface strains although mention is made by Davidenkoff of a design suitable for embedment in concrete. The early BRS gauges clamped onto the material under investigation, strain being transferred via two hardened steel knife edges. The gauge was self-contained and could be unclamped after a test and used elsewhere. For a more permanent installation the clamps holding the wire were fixed to the structure either by welding or by screwing into prepared holes. The accuracy of this gauge decreased a little due to inaccuracies in the gauge length but this was not a significant drawback.

To adapt the clamp-on gauge for embedment was technically feasible but normally too costly to consider. The BRS clamp-on gauge cost £50 each in 1955, and with the necessary modification would have been considerably more. In 1958 Potocki[5] developed a very inexpensive embedment gauge which was subsequently used in large numbers for measuring strains in road slabs and bridge structures. The gauge used a 'Perspex' tube to enclose the wire which meant that considerable care had to be taken when installing it.

As the advantages of the vibrating wire principle became more apparent, other users started to design and make their own gauges. The Central Electricity Generating Board designed a gauge which was used for measuring strains in the biological shields of a nuclear reactor in 1958. At about the same time the Mining Research Establishment designed a gauge which they used for making measurements in mine shafts and underground roadway linings. As the uses increased, gauges previously 'home made' by research organisations or universities, were being made by commercial concerns. In the UK, their manufacture was mainly by instrument makers who used the original designs often without consideration of the various applications for the gauges or of the environment in which they had to operate. The calibration data, supplied with the gauges was assumed to be applicable to any material, which, as will be seen later, was not true. As a consequence, the gauge failure rates were high and it was not until companies who specialised in strain gauges took over the manufacturing that small but significant design changes improved the overall gauge performance.

DESIGN

The strain experienced by the gauge changes the tension of the steel wire and therefore, its natural frequency. The law connecting tension with frequency is

$$f = \frac{1}{2l} \sqrt{\frac{TG}{m}} \qquad (1)$$

where f is the natural frequency (Hz), l is the length of the wire between anchorages (m), T is its tension (Kg), m is its mass per unit length (Kg/m) and G is the gravitational constant (m/sec^2). Equation (1) can be rewritten to give the relationship between strain (ε) and frequency

$$\varepsilon = Qf^2 \qquad (2)$$

where Q is the gauge constant $4l^2m/EGA$ (sec^2), A is the cross section of the wire (m^2) and E is Young's Modulus for steel (N/m^2).

Figure 2 shows a design of a gauge for embedment in concrete where the wire and magnet are contained within a tube and housing. Its design must take into account the material in which it is to be used, if the gauge is stiffer than the matrix then the indicated strains will be lower than they should be and vice versa. The theoretical errors between true and indicated strain for

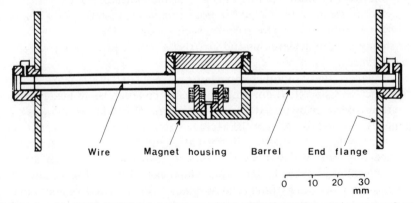

Wire   Magnet housing   Barrel   End flange

0   10   20   30
mm

FIG. 2.   Sectional drawing of an assembled vibrating wire strain gauge for embedment. Courtesy of Strainstall Ltd.

gauges of different stiffnesses have been derived by Loh[6] and this relationship is given by

$$\text{error} = \frac{k(1 - M)}{1 + Mk} \tag{3}$$

where M is the gauge to embedment material modular ratio, E_g/E_e, k the value $[(1 - v^2)\pi]/\{2l/[R - \pi(1 - v^2)]\}$, l/R is the length to barrel radius ratio of the gauge and v is Poisson's ratio.

For a length to radius ratio of between 20 and 200 and a Poisson's ratio of 0·18, the errors resulting from a modulus mismatch are shown in Fig. 3.

An example of the effect of a modulus mismatch can be seen from some results by Browne and McCurrich[7] where similar gauges were cast into a concrete mix and a resin. The concrete had a modulus of 24 000 N/mm² and the resin 8500 N/mm². The gauge in question had a stiffness of 33 000 N/mm². Assuming an l/R ratio of 50 and referring to Fig. 3, the two conditions would give errors of -1% and -9% respectively. The two gauge factors determined by Browne were $3·38 \times 10^{-3}$ and $3·08 \times 10^{-3}$, a difference of 9%.

If a perfect match of moduli cannot be achieved, then the gauge should have a lower stiffness than the concrete so that the error and change in rate of error with mismatch is least. Mature concrete has a modulus of between 28 000 N/mm² and 42 000 N/mm², that for younger concretes is somewhat less. A convenient design aim for the modulus of the gauge would be perhaps about 28 000 N/mm². Using such a gauge with an l/R ratio of 50 in high strength mature concrete, the modular ratio (M) would be, say, 0·65 and the error in strain reading only 1% of the indicated value.

As a surface mounted gauge will not be replacing material, its stiffness will be negligible. Usually the barrel of such a gauge will have a bellows unit or sliding joint to achieve this and give the necessary protection from the environment.

The gauge length is another important parameter in the design of a gauge for concrete. Concrete is not a homogeneous material, being made up of aggregate particles dispersed in a cement matrix of a different stiffness. A gauge is required to measure the average strain, so the ratio of aggregate particle size to gauge length is important. A number of researchers have carried out tests to determine the possible errors involved when using different ratios of gauge length to maximum aggegate size, Fig. 4 shows a relationship obtained by Binns and Mygind.[8] A gauge length of about five times the largest aggregate size is generally considered suitable, so for concrete with 20 mm aggregate the minimum gauge length should be

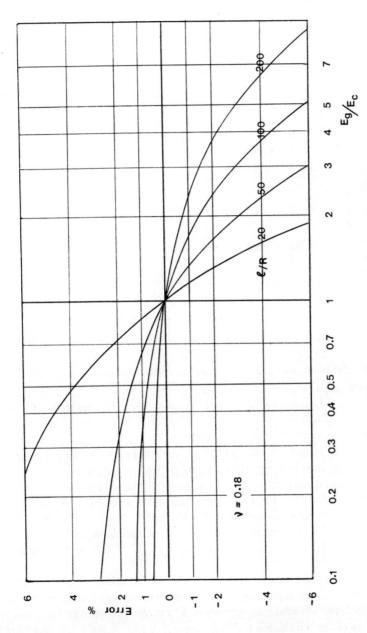

FIG. 3. Theoretical error in strain from embedment gauges due to mismatch of moduli.

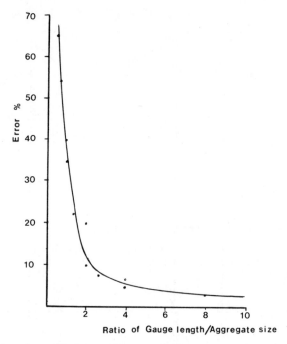

FIG. 4.　Error in strain due to different gauge length to aggregate size ratios.

100 mm. This requirement is applicable to both surface and embedment gauges if used with concrete. If the gauge is used for other structural components, i.e. steel sections or reinforcing steel, then a variety of gauge lengths are possible.

The gauge length choice is also dictated by gauge sensitivity and signal quality. The gauge constant, Q, is proportional to gauge length squared and it can be seen from Fig. 5 that a gauge with a length below 50 mm has a very low gauge constant and hence would have a very low sensitivity. Gauges with lengths as low as 25 mm have been made but it is normal for such gauges to have the wire loaded with a small weight at its centre. This changes eqn. (1) and means that each gauge may have to be calibrated separately. The design is possibly more suitable for external clamp-on types.

The wire, when plucked, needs to be displaced sufficiently so that it will vibrate for a time long enough to enable its frequency to be measured. Therefore, the size of the magnet and coil, i.e. the 'plucking power' and the diameter of the wire needs to be optimised. A coil between 100 and 150 Ω

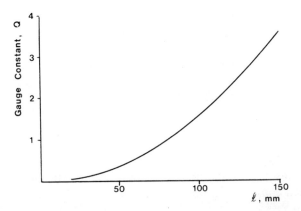

FIG. 5. Relationship between gauge constant and gauge length.

resistance, plucked with a voltage of 50 to 100 volts and a wire diameter of 0·25 mm are normal choices. The magnet and coil require a suitable housing which should be as small as possible as it will produce an unavoidable local stiffening. This housing will affect the gauge factor and is discussed in the section dealing with calibration.

The environment in which the gauge has to operate dictates other design requirements. It will have to resist the ingress of moisture at pore pressures; it will have to operate at above ambient temperatures for long periods and possibly under other hostile conditions such as neutron or gamma radiation. Consequently, the material of the gauge components and their assembly must be carefully considered.

MANUFACTURE

The accuracy and long-term stability of the gauge will depend on the quality of its manufacture. Poor dimensional accuracy in some of the components will result in large variations in the gauge factor. Poor attention to sealing will allow moisture into the gauge which will shorten its active life. Consequently, a high standard of assembly and testing is necessary. The methods adopted by each manufacturer are influenced by his own judgement and by a potential customer's requirements.

Apart from the obvious checking of the tolerances of the machined parts there are some considerations which are not so obvious. For example, the direction of the coil windings and the north and south poles of the magnet should be checked. This will ensure that the pulse through the coil to pluck the gauge is compatible with the direction of the magnetic field. One manufacturer purges his assembled gauges with dry nitrogen and then, after sealing, tests them under water at a pressure of $1.4\,N/mm^2$ for 2 or 3 min. Attention to this type of detail is very important if the failure rate of gauges is to be kept to acceptable limits.

INSTALLATION

If vibrating wire transducers are required as embedment gauges then a careful and consistent installation procedure is important. The gauge is there to measure strain at a desired position and direction and a good transfer of strain is required. Gauges can be cast into the concrete as received or they can be encapsulated in 'dog bones' of concrete before use (Fig. 6). There are a number of reasons why casting-in a bare gauge is undesirable, the most important is the variation of compaction which can occur around the ends of gauges cast in different directions. Tests have shown that the gauge factor of horizontally and vertically cast gauges can differ by as much as 10 %. The reason is attributed to the trapping of air and grout under the end flanges of the gauge. The design of the end plates of some gauges minimises this effect but a gauge with disc ends, such as that shown in Fig. 6, is more prone to this problem. However, encapsulating the gauge horizontally in a suitable mould before use ensures that it can then be subsequently placed in any direction in the mass and behave consistently. Encapsulation will afford considerable mechanical protection during installation, when others are working in the vicinity of the gauge and during the main concrete pour itself.

The concrete used for the capsule should ideally be identical to that used in the main mix. If, however, this mix uses an aggregate with a particle size greater than $\frac{1}{5}$ of the gauge length, then a special mix with a smaller aggregate will have to be designed. This new mix will have the same aggregate type and be designed to have a similar modulus to that of the main mix.

A number of capsule shapes are used but they should be designed so that flat ends and sharp changes of section are avoided. In order that the capsule can attain its bond with the surrounding concrete, a suitable rough surface finish is necessary. It has been found that an exposed aggregate finish is ideal and easily attained. Therefore, before the gauge is positioned in the capsule mould, a suitable concrete retarder is applied to the inner surfaces

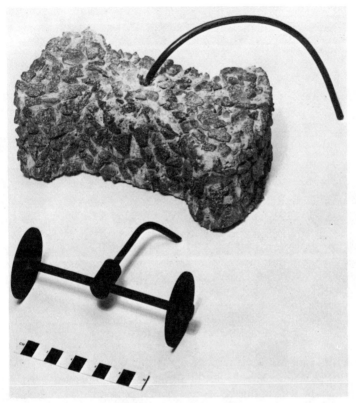

FIG. 6.   Encapsulated vibrating wire strain gauge. Courtesy of the Central Electricity Research Laboratories.

of the mould. The gauge is then fixed in the mould using soft binding wire or nylon cord. After filling the mould with concrete, which should be well compacted, the top surface is sprayed with retarder. The capsule is then kept in a 100 % RH atmosphere for 24 h after which it can be demoulded. By washing it under running water the unhydrated cement at the capsule's surface is removed and an exposed aggregate finish obtained (Fig. 6). The capsule should be stored for a further 28 days at 100 % RH (note that this is not storage under water, storage over water or under damp hessian would be suitable). At the end of 28 days the encapsulated gauge can be stored in air until ready for installation. The encapsulation of gauges should be planned so that the capsules are not older than 3 months when finally concreted in. When fixed in position the capsule should be inspected to

FIG. 7.   Encapsulated strain gauge assembly ready for concreting. Courtesy of the
Central Electricity Research Laboratories.

confirm that its surface is free from any debonding material, i.e. dirt or
grease etc. Capsules should not be wetted before concreting. During the
encapsulation, wire or thin reinforcing rods can be positioned in the mould
which can then be used for fixing the capsule in the formwork. Figure 7
shows a typical installation of a number of encapsulated gauges around a
penetration of a prestressed structure.

Different capsule surface treatments, its age when casting-in and its
moisture history, all affect the gauge factor. The procedure outlined above
is a compromise arrived at after a comprehensive series of tests by
Grainger.[9] However, if a different procedure is adopted it is suggested that
the user standardises it for both calibration tests and for subsequent
casting-in.

CALIBRATION

Calibration tests are necessary to obtain the gauge factor for vibrating wire
strain gauges. This is particularly true for the embedment type, for once the
gauge has been cast into a test sample it can not be used subsequently in the

structure. The accuracy of the measurements from a structure containing these gauges depends then, not only on the quality of the calibration but on the control during the manufacture of the batch of gauges being tested.

Calibration tests are made by casting an encapsulated gauge into a suitable concrete sample and comparing the indicated strain with the measured surface strains during applied load cycling. A recommended minimum size for the test samples would be a 225 mm diameter × 450 mm high cylinder (or 225 × 225 × 450 mm prism) for the 125–140 mm gauges, or 150 mm diam. × 300 mm high cylinder (or 150 × 150 × 300 mm prism) for the smaller gauges. If a cylinder is used, its top surface should be prepared as recommended in the British Standard concrete testing specification.[10] The concrete used for the cylinders or prisms should be the same as that to be used on site.

A convenient way of obtaining the surface strains is with the use of the 'Demec' strain gauge. This is a demountable gauge using pairs of locating discs attached to the sample. A minimum of three load cycles are applied to each sample, the maximum load being determined by the maximum stress expected in a structure. Plots of average surface strain against the differences of the square of the frequency will give the required gauge factor.

The calibrated gauge factor is invariably different from that obtained theoretically. If one considers the construction of the gauge with a stiff housing for the magnet and coil, this is not surprising. The housing forms a stiff inclusion which upsets the strain field and effectively reduces the length over which the gauge operates. Consequently the actual gauge factor is higher than the theoretical one.

Temperature Calibration
It is important to clearly differentiate between the thermal strain of the concrete and the influence of the thermal expansion of the gauge wire. Calibration is required and the procedure normally adopted is given below. The same samples that were used for gauge factor calibration can be used.

Each sample is immersed in a water bath at 20 °C and left until successive strain readings do not change by more than 10×10^{-6} strain units ($10 \, \mu\varepsilon$) in 24 h. It may take several days for moisture equilibrium to be achieved. Readings of both surface strains (Demec strain) and (VW strain) are then recorded, together with the water bath temperature. The temperature of the bath is then raised and after allowing time for the temperature of the sample to stabilise, a further set of readings are taken. The temperature intervals and the range would be determined by the maximum levels expected in the structure. A plot of Demec strains against temperature will give a value of α_c, the coefficient of expansion of the concrete. A plot of VW strain readings against temperature will give a value of $(\alpha_w - \alpha_c)$ from which α_w, the coefficient of expansion of the wire can be obtained.

The thermal coefficients of expansion of different concretes will vary from about 7×10^{-6} to $14 \times 10^{-6}/°C$. That of steel will be about $12 \times 10^{-6}/°C$.

The VW strain, ε_G, obtained using eqn. (2) will consist of three components:

(1)   the thermal strain of the wire, $\alpha_w \times \Delta T$;
(2)   the thermal strain of the concrete, $\alpha_c \times \Delta T$;
(3)   the stress generated strain, σ/E.

Taking compressive strain as positive:

$$\varepsilon_G = \alpha_w \Delta T - \alpha_c \Delta T + \sigma/E \qquad (4)$$

Using a concrete with a limestone aggregate, the calibration tests gave a value of $\alpha_c = 8 \times 10^{-6}/°C$ and $\alpha_w = 12 \times 10^{-6}/°C$.

In a situation where the concrete is free to expand—i.e. no restraint, the stress will be zero and

$$\varepsilon_G' = (12 - 8)\ \mu\varepsilon/°C$$

If, however, some restraint was present then the concrete will be under a thermal stress and

$$\sigma/E = (\varepsilon_G'' - 4)\ \mu\varepsilon/°C$$

Therefore, if the stress raising strain in a structure is required, a correction factor of $(\alpha_c - \alpha_w)/°C$ should be added to the gauge indicated strain.

To measure temperature at the gauge position and to avoid the need for a separate transducer, it is possible to use the plucking coil of the gauge as a resistance thermometer. This is provided with a datum resistance at $20°C$ and a coefficient of resistivity, from which the gauge temperature can be derived. Such coils are wound to a resistance of say, 135 ± 0.2 ohms. The change of resistance per $°C$ for such a coil would be approximately $0.5\ \Omega$.

ACCURACY

From eqn. (2), it can be seen that strain is proportional to the length^2 and the mass/unit length of the wire and inversely proportional to its modulus.

A change of 1 % in each of these parameters would give, at an average
frequency, an error of about 4 %. Although the measurement of frequency
can be obtained to a very high degree of precision, this is pointless if the
variations in the other parameters do not warrant it. The gauge factor
determination, for example, is only as good as the surface strain
measurement. The temperature correction is dependent upon the accuracy
of the temperature measurement. It is therefore unrealistic to expect to
measure strain to an accuracy better than $10\,\mu\varepsilon$ without going into
elaborate methods of gauge testing and selection.

For long term accuracy, the effect of the creep of the wire has to be
considered. In a typical gauge at a frequency of 1000 Hz, the stress in the
wire would be about $550\,N/mm^2$. Small but significant creep would
normally occur at these stresses particularly if the temperature is above
ambient but this problem can be largely eliminated by 'precreeping' the
wire. There have been a number of tests performed which compare the
performance of treated and untreated wire. Figure 8 shows the results of
work by Taylor Woodrow Construction Ltd. In their tests, wire that had

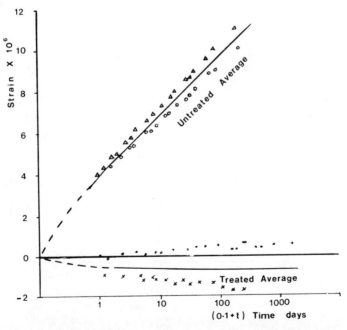

FIG. 8.   Creep of treated and untreated wire at 65 °C under load.

been loaded at 940 N/mm² and kept at 200 °C for 24 h was compared with untreated wire. Tests by the author have also confirmed that creep of treated wire is of no significance and that the majority of creep that does occur is during the first few weeks of a newly tensioned gauge, long before it is installed in a structure.

Strain Range

The range over which the gauge can operate is governed by the stress in the wire at one extreme and by the quality of the signal at the other. Figure 9 shows the relationship for a 125 mm gauge ($Q = 2·4 \times 10^{-3}$). It would be unwise to stress the wire above say 950 N/mm² which would give an upper limit of frequency of a little under 1400 Hz. At the other end of the scale, a poor signal (and hence a loss of accuracy) would occur at frequencies below 500 Hz. For these two limits, the strain span would be 4100 $\mu\varepsilon$. If it is known which direction the strain in the structure will go (tensile or compressive) then use can be made of most of this range. For example, in a prestressed structure, the maximum strain at many places is that due to the prestress and is decreased by loading. A gauge tuned to a high frequency during

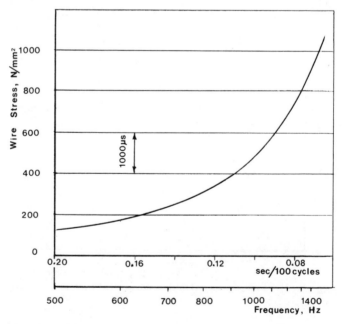

FIG. 9.   Stress and strain range for vibrating wire strain gauges.

manufacture would in such a case have 3000 or 4000 $\mu\varepsilon$ of 'available strain'. More generally though, the frequency specified would be about 1100 Hz, giving a strain range of $\pm 2000 \mu\varepsilon$.

LOGGING

In the early days a disincentive to the use of vibrating wire strain gauges was the difficulty in obtaining the frequency measurement. The early recorders would balance the signal from the active gauge with that from an adjustable control gauge or signal. The two outputs were then compared using Lissajous figures or by 'beats'. These methods were laborious and inaccurate as well as being unsuitable for any sort of logging, where a number of gauges had to be scanned. When the use of the gauges became more widespread, better recording devices were developed usually based on the absolute measurement of frequency. The designs improved but even in the late 1960s the loggers were still very large and it was not until solid state switching and integrated circuits were widely available that the size was reduced to an easily transportable unit. The size of a present day 80 channel logger is about the same as an electric typewriter. These loggers use a frequency meter and either measure frequency directly by counting the number of cycles over one second or obtain the reciprocal frequency by timing a predetermined cycle count (usually 100). The latter method is the more accurate and is normally adopted. The recorder, when started, initiates a pulse to 'pluck' the wire and then, after a suitable delay of a few microseconds, starts the cycle count. It is common for the output wave form to be electronically shaped to a saw-tooth to increase the precision of the count. The recorder can have the ability to pluck the gauges with a range of voltages, so that the larger gauges can be given a stronger pluck.

As the output signal is dependent on the wire cutting the magnetic field of the permanent magnet, the existence of this field is important. On a number of gauges in an early installation, the signal was found to be deteriorating with time. Subsequent investigations found that the plucking pulse was in opposition to the magnet and that every pluck was demagnetising it. The direction of the coil around the magnet is thus of importance. The sequel to this episode was that after reversing the connections and plucking the gauges continuously for a few hours, enough magnetism was restored and the gauges operated successfully for many years.

The mains supply to the recorder should be reasonably clean and free from externally induced spikes. In a recent test a proportion of the gauge

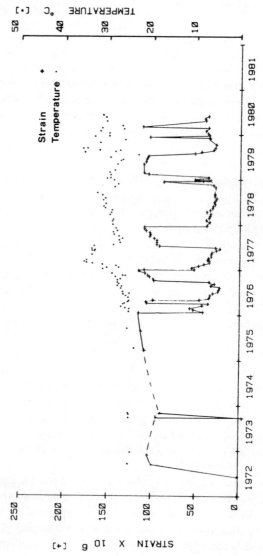

FIG. 10. Automatic plot from strain gauge recorder (Gauge 10).

readings were found to be departing from the expected values. Plots of strain against load for each gauge produced two parallel lines. The explanation of this was found to be random spikes on the mains supply. The recorder counted 99 cycles of the gauge and one spike from the mains. By reducing the time recorded by 1 % and recalculating the strain, all the readings were brought back to one line.

Loggers are produced where a set of gauges are sequentially plucked and the frequency or period measured. An analogue reading is also made of the coil resistance immediately before it is plucked. A typical output from such a logger would include date, time, channel number, resistance, reciprocal frequency and a logger identity character. The latter is necessary when several loggers are used on the same installation. Interfaces are now fitted so that the output can drive, for example, a magnetic tape recorder. This then enables the data to be subsequently processed using a microprocessor.

The author used a logger, a cassette tape recorder and a small micro-processor based computer. The reduced data is stored on floppy discs from where it can be accessed for analysis and automatic plotting (Fig. 10). Where a large number of gauges are used on a structure, with readings taken at regular intervals for perhaps several years, such an automatic data handling system is very cost-effective.

An important difference between recording data from vibrating wire transducers and from those based on an analogue measurement is the formers tolerance to lead lengths and leaks to earth. As the output is frequency dependent, very long lead lengths are possible and any reduction in the signal level can be recovered with suitable amplification. Distances of over 2 km (cable resistance of about 300 Ω) have set no problems. For the same reason poor junctions or high switching resistances or low resistances to earth cause no change in frequency reading. A gauge with a short to earth of 100 Ω still gave a measurable signal and was the same value as that without the short. These are useful bonuses should accidents happen.

APPLICATIONS

In many cases, strain gauges are used to provide data to confirm that a structure is behaving in accordance with design predictions and, in certain cases to provide data from which the ultimate load factor can be derived. The development of the prestressed concrete pressure vessel (PCPV) as a nuclear containment necessitated the construction and testing of large models. The scales of these models were between 1/10th and 1/8th of the full size prototypes. This enabled real concrete to be used for their construction and for embedment gauges to be cast in a number of positions within the concrete section.

One of the models for the first PCPV in the UK, that for the Oldbury Power Station, had 360 VW strain gauges cast in the concrete during construction. The transducer selected had a gauge length of 96 mm and was encapsulated as described earlier. Readings were taken during the prestressing operations and then during the pressure testing of the vessel.[11] Further tests were then undertaken with thermal gradients applied under long-term loading when the measurement of creep strain was of importance. Finally, the model was tested to the ultimate and the gauges gave data of the crack pattern that had formed at this load. Figure 11 shows a strain response at two cracked zones, one where a crack passes through a gauge and the other where the crack is adjacent to it.

The sequel to these tests was the building and proof pressure testing of the full size vessels. The first of the two built had 350 gauges incorporated in the concrete, their gauge length being 133 mm. Readings from the gauges were taken during prestressing and pressure testing and have been regularly taken during the 14 years (to date) of the commercial operation of the station. The pressure test mentioned is a statutory one and the VW strain gauge results provided evidence that the vessel, from the structural point of view, was safe to operate.

The readings from the model and the full size prototype gave a unique opportunity to confirm the original design predictions and enable analytical methods to be refined for subsequent designs.

It is of interest to note from the strain history shown in Fig. 12, the significant portion of total strain that is attributed to creep strain. In order to determine the variation of creep strain with stress, temperature and time, a large programme of creep tests was undertaken. In these tests, strain measurements were required from a large number of samples under different loading regimes for several years. It would have been a daunting task if all the readings had to be taken using, for example, a demountable extensometer. In some cases, where triaxial loading was applied to the sample, there was no opportunity to measure surface strains and a remote reading transducer was essential. The VW strain gauge proved an excellent method of obtaining the data required with the minimum of effort.

Tests and installations such as those just described have been made for all the 18 nuclear concrete pressure vessels operating or under construction in the UK.

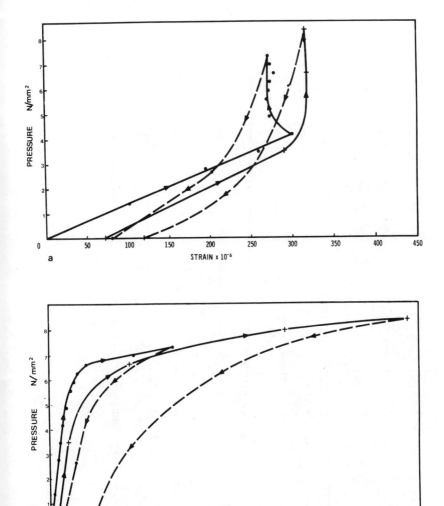

FIG. 11. Influence of cracking on strain gauge behaviour: (a) plot where gauge is adjacent to a crack; (b) plot where the crack passes through the gauge. Courtesy of the Central Electricity Research Laboratories.

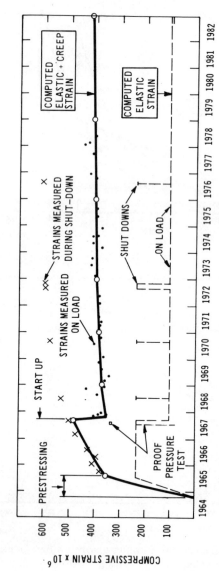

Fig. 12.   Strain history of a gauge from an operational concrete reactor vessel.

As a founder developer of the VW strain gauge, the Building Research Establishment have made wide use of these gauges. Stevens and Corson describe four applications in one of their papers.[12] One structure described was a multi-storey office building where strains were measured over an $11\frac{1}{2}$ year period. The measured contractions of the prestressed concrete beams due to creep and shrinkage compared well with those estimated from the recommendations of the (then) Code of Practice (CP 115).

Gauges have been used by the Transport and Road Research Laboratories for the study of the behaviour of road bridges. Apart from information gained from direct measurement of strain an interesting method of determining stresses has been developed by Tyler.[13] Hydraulic creep rigs are used to load concrete samples cast at the same time as the structure being investigated. As the structure is being loaded, say, by prestressing, the concrete sample is also loaded, the load being determined by matching the strain in the sample with that in the structure. As the strain in the structure changes due to creep, shrinkage or stress redistribution, so the load on the sample is adjusted to keep the strains in step.

With the use of concrete for new mine shafts and pit bottom roadway linings, it was necessary to have a greater knowledge of the loads to which these were subjected. The Mining Research Establishment undertook a large programme of gauge installation and long term monitoring to gain this information.[14] The work started in two shafts at the Wolstanston Colliery in Staffordshire and in a difficult geological area of a new sinking at Abernant in South Wales. The gauges were generally cast in pairs, the distance separating them being dependent on the thickness of the section. Thus strains near the inner and outer surfaces of the concrete were measured, from which bending and direct strains could be obtained. Data obtained from such installations were able to influence the design of future construction in such factors as the best shape of lining for a particular location and the amount and position of the reinforcing steel. The use of vibrating wire strain gauges in such locations several thousand feet underground is a good example of the hostile conditions under which they can operate.

CONCLUDING REMARKS

The last few pages have given examples of the use that the vibrating wire strain gauge has been put to in the UK. There are many applications around the world not mentioned but of no less significance, as indeed has been the gauge's development in, for example France and Germany. The chapter has dealt exclusively with the strain use of the vibrating wire principle. This principle has, however, been used for many other transducers; the measurement of deflection; load; inclination and pore water pressure, for example.

There is little doubt that the development work which has produced the VW gauge has contributed to engineers being able to obtain greater understanding of the actual behaviour of structures in the field. Its reliability over many years has been proven and has meant that not only the elastic behaviour of a structure can be confirmed but of possibly greater value, the long-term shrinkage and creep can be studied. By instigating a monitoring programme, structures of importance—road bridges, nuclear containments etc., can be watched to detect any departure from the limits imposed by the designers, a practice that can only improve the knowledge and safety of all structures.

REFERENCES

1. DAVIDENKOFF, N., The vibrating wire method of measuring deformations, *Zh prikl. Fiziki* (*Journal of Applied Physics*), Leningrad, 1928, **5**.
2. JERRETT, R. S., The acoustic strain gauge, *J. Sci. Inst.*, 1945, **22**.
3. MAINSTONE, R. J., Vibrating wire strain gauge for long term tests on structures, *Engineering*, July 1953.
4. WHIFFIN, A. C., Some special techniques developed at the Road Research Laboratory for testing roads and other structures, *RILEM Symp. on the observation of structures*, Lisbon, 1955.
5. POTOCKI, F. P., Vibrating wire strain gauge for long term internal measurements in concrete, *The Engineer*, Dec. 1958.
6. LOH, Y. C., Internal stress gauge for cementitious materials, *Proc. Soc. Exp. Stress Analysis*, 1954, **11**.
7. BROWNE, R. D. and McCURRICH, L. H., Measurement of strain in concrete pressure vessels, *Inst. Civ. Eng. Conf. on PCPVs*, London, 1967.
8. BINNS, R. D. and MYGIND, H. S., The use of electrical resistance strain gauges and the effect of aggregate size on gauge length in connection with the testing of concrete, *Mag. Conc. Res.*, 1949, **1**, pp. 35–9.
9. GRAINGER, B. N., Encapsulation of vibrating wire strain gauges. The effect of capsule age and treatment, *CERL Report RD/L/N50/78*, Aug. 1978.
10. BRITISH STANDARD, Methods of testing concrete, BS 1881 Part 5, 1970.
11. HORNBY, I. W., VERDON, G. F. and WONG, Y. C., Elastic tests on a model of the Oldbury Nuclear Station PCPV, *Proc. Inst. Civ. Eng.*, July 1966.
12. STEVENS, R. F. and CORSON, R. H., Strains in some prestressed concrete structures. *Inst. Civ. Eng. Conf. on Stresses in Service*, London, 1966.
13. TYLER, R. G., Full scale tests on the Mancunian Way elevated road, *Inst. Civ. Eng., Proc.*, Paper 7196S, May 1969.
14. SHEPHERD, R. and WILSON, A. H., The measurement of strain in concrete shaft and roadway linings, *Trans. Ins. Mining Eng.*, 1960, **110**.

INDEX